About the Size of It

Warwick Cairns is also the author of *How to Live Dangerously*, a book that re-assesses the real risks of modern-day life and urges us to worry a whole lot less. He lives in Windsor with his wife and two daughters.

Praise for About the Size of It

'Fun and fascinating – the secrets and tricks of how we measure the world around us. It answered a lot of questions for someone born in the seventies, I can tell you' **Conn Iggulden**

'The metric system was born of French revolution, but though Napoleon lauded it, he did not use it himself and predicted it would be a stumbling block for generations. Warwick Cairns, in his deliciously ironic history of traditional measurements, agrees. He maintains that while the International System of Units is vital to science, everyday measures for length, volume and weight should relate to the "dimensions, appetites and purposes of ordinary people". After ten years conducting opinion polls, he really does know what we think'

New Scientist

'Warwick Cairns's superb work makes a seemingly complex subject simple to understand. His direct, engaging conversational prose is a delight to read' **Andrew Roberts**

'Warwick Cairns appeared in court as an expert witness for market traders determined to continue selling their bananas by the pound. It turns out that consumers remain attached to old-fashioned measuring systems for more than merely jingoistic reasons. Rather, it is because they work, based as they are on common-sense approaches developed over generations . . . If you're easily wound up about metric measures having taken over from imperial, then this passionate, entertaining defence of traditional measuring systems will surely provide solace'

Oxford Times

WARWICK CAIRNS

About the Size of It

PAN BOOKS

First published 2007 by Macmillan

This edition published 2015 by Pan Books
an imprint of Pan Macmillan, a division of Macmillan Publishers Limited
Pan Macmillan, 20 New Wharf Road, London N1 9RR
Basingstoke and Oxford
Associated companies throughout the world
www.panmacmillan.com

ISBN 978-1-4472-9147-3

Visit **www.panmacmillan.com** to read more about all our books
and to buy them. You will also find features, author interviews and
news of any author events, and you can sign up for e-newsletters
so that you're always first to hear about our new releases.

This book is for my wife Susan
and my daughters Alice and Lucy

Out of the crooked timber of humanity,
no straight thing was ever made.

Immanuel Kant

Contents

Introduction

This is a serious book about weights and measures.

You can take that as a warning, or else you can take it as an invitation.

Either way, that's what it is, and it's what you're going to get if you decide to read on.

In these pages you will learn important things. At times, you will look at the world in ways that your earliest ancestors saw it, and find their understanding not so very different from your own. You will also see how best to slice up a Bakewell tart without making a complete and utter khazi of things, and learn how to use your thumb as an aid to carpentry. As you read on, you will discover one of the most powerful 'trade secrets' of professional photographers, and find out what happens when, or if, you drop a large stone on your hand. After reading this book, you will find yourself more than able to hold your own in any discussion on the relative merits of vegetables and cereals as aids to the correct sizing of shoes – even if that discussion is with the lady from your local shoe shop. And – most of all – you will understand the one great unwritten, unspoken, unacknowledged Principle of Measurement.

But I'll warn you now that it does go round the houses a bit, this book, in getting to where it ends up. Which I did, in fact, in coming to write it, and in coming to set down here what I now know.

How I came to it was when the British government introduced laws, a while back, to convert the people from the characterful (you could say 'bonkers') traditional system they had used in one form or another since – well, since before anyone could remember – to the rational (some might say 'soulless') metric system. At this time, people had been allowed to use the metric system for trade and all sorts of things for a hundred years or more, but few of them seemed to be doing so, for reasons best known to themselves.

I was called in to commission a series of opinion polls to see what the public thought about the change to metric, and how – or whether – what they thought about it changed over the years. I did this for almost ten years. And then, because no one else had done this sort of research into the subject, when the first cases came to court and market traders were hauled up before the bench for selling bananas by the pound, I was brought in as an 'expert witness' to explain why they did what they did, and why they should be allowed to continue doing so. As those cases progressed from local courts to the High Court in London, I spent long hours in conclave with these traders' defence lawyers, and with an Oxford psychologist, trying to work out not just what the results of the research said, but what they meant, and what was going on behind them in people's minds to make them think and behave the way they did.

To sum up ten years of research and seven thousand interviews in a very few words, the gist of it was that people didn't have a very high opinion of the change, if the truth be told; and over the years their opinions changed very little indeed. If anything, they became slightly less happy about it, rather than more so.

Which you could sort of understand, if people as a whole, or just British people, were none too keen on change in general; but over the exact same space of time these same people were happily embracing all sorts of other changes in their lives without a second thought. They ditched their clunky old video recorders and bought DVDs in their millions. They took their big fat cathode-ray televisions down to the dump to make space for plasma and LCD flatscreens. They fitted microwave ovens and dishwashers in their kitchens, and changed their hairstyles and their cars, their mobile telephones and the cut of their trousers – and just about everything else you could care to mention – in order to keep up with the times. And they did it because they wanted to, and no one had to pass any laws to force them to do it.

But when it came to measurements, things were altogether different. And as it happens, things always have been different, right throughout history, where measurements are concerned. Throughout history, whenever a government has decided that its people might be better served by swapping their own traditional system for a shiny new one, those people have been less than grateful, and less than enthusiastic, and often surprisingly stubborn in their resistance.

In France, years and years after the Revolution, the nation that had happily guillotined its king and done away with centuries of tradition remained so hostile to the *metre* and the *kilogramme*, and so fond of the *pied de Roi* and the *livre de marc*, that the Emperor Napoleon eventually got fed up and passed laws allowing them to change to the old *mesures usuelles*. Which they did, pretty comprehensively. It was only after Napoleon went, and a new government of hardcore modernizers got in, that the new system came back again; and it was only when the full force of the criminal law came down on them that the people grudgingly accepted that the game was up, and did what they were told.

Which seems to have been the case all round the world.

This all makes you wonder what's going on here. It makes you wonder, for example, whether there's something in the nature of people that keeps them wedded to certain ways of weighing and measuring, even as the world around them changes; or else it makes you wonder whether there's something in those ways of weighing and measuring that makes people want to keep on using them.

And that, pretty much, is what this book is about.

And so now it's probably a good time to step back from laws and trials and opinion polls and historical precedents, and to move swiftly on to the great unwritten Principle of Measurement, and then to the uses of wellington boots.

The Principle of Repeated Bodges

Why people measure things the way they do isn't something that most of us agonize over too long or too often. Most of the time we just do what needs to be done without thinking much about it.

But if you do want to know, and if you consult the textbooks, you will see that most of the reasons written down there are to do with history, or with science, or the history of science, or even, perhaps, the science of history.

This king, seated on his draughty throne back in the twelfth century, decreed that the length of his royal appendage should be set as the standard of measure for such-and-such for all eternity, and had it inscribed with quill pens on vellum scrolls, now slowly mouldering in some vault in the British Museum, or the cellar of the House of Lords or somewhere. That white-coated scientist, hunched over his test-tube rack in the dead of night, chewing distractedly at the end of his pencil, discovered the Theory of Such-and-Such, and in gratitude the unit of its measurement was named after him. And so on, and so forth.

But scientists and kings, on the whole, make up a very small part of the population, and the measurements they make are far fewer in number, in the grand scheme of things, than the sorts of measurements that normal people ordinarily make going about the business of their daily lives.

The sorts of measurements that normal people ordinarily make going about their daily lives are guided by one great unwritten, unspoken, unacknowledged Principle of Measurement. This is something that's a lot less grand and exalted in reality than it sounds when you say it, but it's none the less important for that. Expressed simply, it's the fact that most people – how shall I put this? – it's simply the fact that most people can't always be bothered to do things properly.

Always doing things properly can be a bit of a fag, you see; and always doing measuring properly involves carrying all sorts of measuring equipment around with you – rulers and balances and measuring jugs and the like – just on the off-chance that you might be called upon to use them. This is fair enough for the measuring obsessives of this world, but the rest of us tend, instead, to go in for bodges, cheats, compromises, estimates and rules of thumb.

Some of these improvised measurements sort of work, and some of them don't. Some of them turn out to be really rather good, useful and reasonably accurate – good enough, in fact, to consider using them again, and to recommend them to others – while others turn out to be a lot less so.

When you keep on doing this, day after day and year after year, and when you keep on doing it for generation after generation and for century after century, what tends to happen is that the best and most useful ones stick and get handed down and passed on again and again and again, while the others fall by the wayside until you find, almost by accident, and almost without planning it, that you've ended up with a system.

That's how the Principle works, and not just for measurement: if it's simpler and more convenient to do something in the 'wrong' way than in the 'right' way, then people will tend to do it; and if enough people do it in enough 'wrong' ways over a long enough period of time, then one of those 'wrong' ways often ends up becoming the 'right' way. And in this way systems evolve towards the way normal people really are, and the needs and uses normal people really have; and away from the grand theories and structures created for them by their 'betters'.

We'll call this the Principle of Repeated Bodges.

The Builder's Measuring Boot

Let's have a go at a bodge, an amateur measurement, to see how it's done. And let's start with you, and where you are right now. Right now, let's measure up the place you're in.

You have two choices here.

One – if it's easy, and practical, and if you can be bothered, you can somehow get hold of a tape measure or other measuring instrument, and use it in the proper way.

Or two – you improvise with whatever tools you happen to have to hand. And of all the tools you have to hand, right now, for measuring out the place you're in, the best and most useful for improvising with are your own feet. They are natural, intuitive things to use – and, as we'll see in a little while, they're surprisingly accurate.

To get a sense of how common it is to measure with your feet, it's instructive to take a walk down to your nearest building site. There you will see people who do this for a living – bricklayers, carpenters, electricians and plumbers – translating architects' millimetre-perfect blue-prints into solid physical reality, both with tape measures and rulers, when they have them, and also – when they don't – using their own builders' measuring boots.

A measuring boot

All they do – and all you need to do to measure up the place you're in – is to put the back of your heel against where you want to start and walk carefully forward, heel to toe, until you get to the other side, and count the number of steps.

So now you've got yourself a unit, and a measuring instrument you can carry about with you wherever you go. You can even give it a name: depending on who you are, and where you live, and what language you speak, you can call it a *foot*, or a *fod*, or a *pied* or a *pie* or a *fuss*, or even, if you live further afield, a *shaku* or *kanejaku*.

Men, Women and the 'Official Boot'

Measuring things with your own feet, for your own purposes, is a useful and natural way to go about things, but what happens when you want to compare your measurement with someone else's?

We all know that different people have different-sized feet – but what matters is how different, and how much that difference matters, for practical purposes, and how easy or difficult it is to iron out any differences that do matter.

'Different people have different-sized feet'

Now, as it happens, people's feet don't vary quite as much as you think. The chances are, if you are a man, then the shoes you're wearing now (or, if you are a boy, the shoes you will wear when you grow up) will be one of just three sizes: an eight, nine or ten in British sizes. Three-quarters of all British men fall into that narrow range, with almost all the rest being no more than one size on either side. Actually, before I go on here it's worth pointing out that different nations do shoe sizing differently. The Americans use the same sizes as the British but number them differently, so that our ten is their eleven. The continental Europeans use a different system altogether, such that a UK ten is a forty-four. There are reasons for this, and I explain them later, but for now let's just stick to the British version, for no other reason than it's the one most people reading this edition of the book will be familiar with.

The difference between an eight and a ten is about a finger's breadth, which, for rough measurements like sizing up a room, makes very little difference indeed.

If you are a woman, your feet will most likely be smaller; but again, the chances are that your shoes will be in one of just three sizes – a UK five, six or seven.

So how big a difference does the difference between men and women make to measuring things?

Well, if you take the biggest 'normal' men's size – a ten – and the smallest 'normal' women's size – a five – the ratio between them is roughly five to six.

Which is to say, if a small-footed woman walks six steps she will have covered the same distance as a big-footed man covers in five.

Whether that counts as a big difference or a small one depends on what you're measuring and how accurate you need to be. But throughout history most societies have tried to impose some degree of consistency and uniformity on things by coming up with an 'official' foot.

The current 'official' foot in use in Britain and the USA is the size of the sole of a man's British size-ten shoe (or a size-nine workman's boot).

Most other countries at various times have had their own foot-based measures, and, as you would imagine for different measures that have developed independently in widely different environments, they have varied in size. But not by nearly as much as you might think.

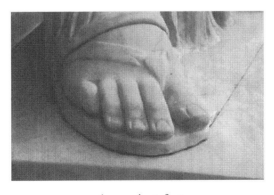

An ancient foot

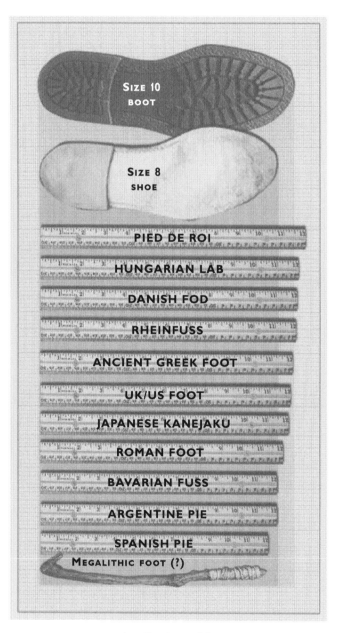

Size 10
BOOT

Size 8
SHOE

PIED DE ROI

HUNGARIAN LAB

DANISH FOD

RHEINFUSS

ANCIENT GREEK FOOT

UK/US FOOT

JAPANESE KANEJAKU

ROMAN FOOT

BAVARIAN FUSS

ARGENTINE PIE

SPANISH PIE

MEGALITHIC FOOT (?)

A collection of feet

The Builder's Measuring Boot

If you set out all of the different 'foot' units, from the versions used by the ancient Greeks and Romans to the versions developed as far afield as Spain, Argentina, Hungary and Japan, almost all of them fall within the narrow range of shoe sizes worn by most adult males today.

And what little variation there is seems often – though not always – to be related to the climate, which in turn relates to whether men at work wear thicker, warmer work-boots or whether they wear lighter footwear.

So the Hungarians, the Danes and the north Germans developed feet in the region of a ten-and-a-half to eleven shoe (or a nine-and-a-half to ten boot), whilst the Spanish and the south Germans developed foot measures at the smaller end of range.

The thirty-centimetre rulers commonly sold in metric countries that don't officially have a foot-based measurement are actually the equivalent of a size nine.

Of all of the different foot measures used over the years, only two fall any appreciable distance outside the average range of modern men's shoe sizes; and one of those might not have even existed.

The biggest, which is the equivalent to at least a size twelve, is the Parisian *pied de roi*, or 'king's foot'. There may be something of a clue to its rather whopping size in the name. Rather than being an average workman's boot, this measure was altogether less plebeian. Perhaps there was something of a fashion for fancy footwear in the French court at the time the measure was standardized, or perhaps there was an intention to impress the peasantry with the size and puissance of His Majesty's person; either way, this measure is seldom, if ever, used today.

A Megalithic foot

The smallest – and also the least well documented – is the so-called *Megalithic foot*. Based on measurements of stone circles and other ancient monuments carried out by Alexander Thom, a professor of engineering at Oxford University, there is a suggestion that many of the European tribes of some six thousand years ago used a system of measures that might or might not have included a 'foot' that translates roughly to a modern six-and-a-half or seven. This would be the equivalent of a seven-and-a-half or eight foot bare, or wearing thin animal-skin moccasins; and this, in turn, would seem to be just about within the bounds of 'normal' by today's standards, if a little on the small side. However, the jury is still out on Thom's conclusions, so we'd need to proceed with a little caution here.

The conclusion of all of this, though, is that for all the differences in the sizes and styles of various countries' shoes over the past few thousand years, the principle is more or less universal: when it comes to measuring short and medium distances, the most widely used instruments, and units, are the ones on the ends of your legs.

The Carpenter's Hand Tools

Measuring out distances with your feet is a good improvisation, but it does have its limitations.

You can measure a room, or a field, or the foundation ditch for a wall, but unless you're Spiderman you won't be able to walk up your wall once you've built it, to see how high it is.

Actually, that's not strictly, one-hundred per cent true. By lying on your back you can sort of get about three or four steps up a wall (I managed three and a bit). It does look and feel a bit silly, though, and depending on where you are you could get rather dirty doing it.

Or you could take your boots off and stand in the mud in your socks while you walk them up the wall with your hands – that way you'd be able to get in six, seven or even eight 'steps' before you'd have to give up. This is assuming that you've managed to withstand the mockery of onlookers while you're at it.

'So you want me to lay on my back and walk up the wall . . . ?'

The other thing you could possibly do is to take hold of the thing that's standing up, lay it down, walk along it and then stand it back up again; but for a lot of things – and not just walls – that's not always a very practical thing to do.

What people all over the world have done, in the absence of rulers and tape measures, is to use what for most of us are the body's second most popular measuring devices: the hands.

For nomadic herders, hill farmers and others whose concerns are less to do with the width of rooms and enclosures and more to do with the height of plants and livestock, hands have long been more important than feet for measuring.

To measure your horse, cow, or other upright object, you start at the hoof or base, with your hands sideways across, and 'walk' them up, hand over hand at the widest part, which is from the joint where the thumb meets the hand to the crease of the palm on the other side. Because animals can change the height at which they hold their heads depending on their mood or whim, you stop not at the ears or the top of the head, but at the withers, which is the highest part of the back around the top of the shoulder.

Like feet, hands vary in size; but also like feet, the variations in the sizes of individual hands tend to be clustered fairly tightly around a widely shared 'average' value for men, and similarly for women.

And, as for feet, many societies throughout history have had a hand-sized measuring unit. The Anglo-American system has set a value for a standard or average *hand*, mainly used for measuring horses nowadays, and we'll be doing a little experiment later with a strip of paper in which you can – if it takes your fancy – work out how big it ought to be, from first principles; but there are one or two other things we need to clear up first before we get to that.

You may or may not have measured a horse, but one occasion on which you will have come across multiples of the 'official' hand – even though you may not have been aware of it – is if you have ever used the sort of retractable tape measure sold at builders' merchants and DIY stores.

The reason you may not have been aware of it is that the measure comes as a cryptic little mark which people in some countries have little or no use for, and so never even notice. It all depends on what they use to build houses with where you come from. If you live in a country where bricks and stone are the main construction materials, then the use of hands (and feet) isn't particularly important in judging height, because once you know the height of one course of bricks you can tot up the height of any wall you like.

Built to the scale of the human hand: how bricks work

In general, bricks are the size and shape they are so that you can work with them easily, and so that you can easily estimate the dimensions of the wall you've just built, without carting all sorts of rulers and tape measures around with you. There are slight variations in size from country to country, but in Britain the width of a brick seen from the top, or turned end-on, is one hand's breadth, because your hand is what you pick it up with.

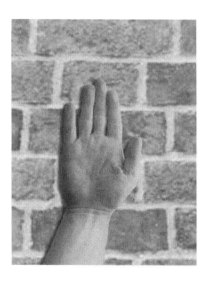

The length of the long face, or 'stretcher', and the height of it, are such that, when laid in a wall with a layer of mortar in between, four bricks along measures three feet, and four bricks up measures a foot.

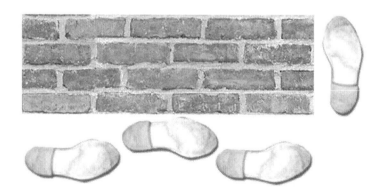

There's a slightly more complicated way of laying bricks, known as 'English bond', or 'garden-wall bond', with bricks laid alternately lengthways and crossways. In that style, a row of eight bricks works out at five feet in length.

Counting bricks is all very well, but if you live in a country such as the USA where timber is more widely used, the wooden studwork with which walls are constructed is designed around multiples of hands, rather than a number of feet, so that once you know where one stud is to knock a nail in, you can find all of the others – even if they are hidden behind plaster – simply by 'walking' up four palms' breadths with your hands.

And for this reason, builders' tape measures have little diamond-shaped marks spaced out every four standard hands.

Rules of Thumb

Feet are all very well, and hands are all very well, but a lot of the things we deal with in day-to-day life don't come in exact multiples of either; and nor do the differences between things we might be able to measure roughly with our hands or feet. Your room, or your mud hut, might be a fraction of a foot bigger or smaller than your neighbour's. Your sheep may be a fraction of a hand shorter, or you may want to knock not one but two or three nails, evenly spaced, into the stud concealed beneath the plasterwork. So again – assuming you don't always have a ruler or a tape measure on you – what are you going to use?

Your hands and your feet are the most mobile and flexible 'tools' you have: it would help if there were some subdivision of those, like your fingers or your thumbs or your toes. Realistically, you aren't going to take your shoes off – and even if you did, your toes don't move about anywhere near as much as you'd need them to if you were going to measure with them. Your fingers and thumbs do, however, and of these, the thumbs – which you can stretch right out at right-angles to your palm – are probably the easiest to use.

A thumb

This seems to be what people have done throughout history, and what they continue to do – to measure a distance smaller than the width of a palm using the thickness of the thumb. So a carpenter, for example, will knock a nail in, and then, steadying the next nail with his thumb, knock that one in a 'standard' thumb-distance higher or lower. Plumbers have a use for thumbs, too – they find them handy for sticking down the ends of pipes while they're working, to temporarily stop the water flow. This is why many pipes around the world tend to be a standard thumb's thickness in diameter.

Thumb nails

Almost every European country has a word in its language for a small standard measure which is the same thickness as a thumb (or the space marked out by a nail knocked into a plank on either side of a thumb), and which is described by either the same word as 'thumb', or one that derives from the same root. This is true both for all of the Latin-based languages and also for most of the Germanic ones (English being one of the few exceptions). The Dutch have *duim*, the Swedes *tum*, the French *pouce*, the Italians *pollice*, the Spanish *pulgada* (from *pulgar* for thumb) and so on and so forth.

So now we have three different measures for length – feet, hands and pouces, or thumbs; but at the moment we have little idea of the relationship, if any, that exists between them. This is something we need to look into.

The DIY Ruler

At this stage, we appear to have three different ways of measuring what is, in essence, the same thing.

We've got the length of a foot, for measuring horizontally, the width of a hand for measuring vertically and the thickness of a thumb for smaller measures or to make up the difference when something's not an exact number of feet or hands long, wide or high.

We've covered the fact that there are standard values for these measurements, based on the average range into which the majority of adult males fall. But what we don't know, at the moment, is the relationship between the three units, or the answer to why the length of a plank of wood, for example, should be described in feet if it's lying down, or in hands if it's standing up. And what if it's leaning against a wall at a 45-degree angle? Should that be hands, or feet? And what about thumbs?

So before we go any further we need to resolve this; and the way we are going to do that is with a very simple experiment (or practical exercise, if experiment sounds too grand).

You will need a sheet of paper at least as long as the sole of your shoe, a pair of scissors and a pen or pencil.

The first thing we need to establish is the relationship between hands and feet. Cut a strip of paper about as wide as a ruler, and the same length as your shoe. Then, using the hand-measuring method, from thumb joint to palm crease, see how many times the width of your hand fits into the length of your foot, and fold your strip of paper into that many pieces.

Unless you are very unusual indeed, or unless you are wearing spectacularly pointy shoes, you will find that your hand will fit almost exactly three times into the length of your sole, and your paper will now be folded into three equal parts. This is one of the basic rules of human proportion, like the fact that the distance from your kneecap to the ground is twice the length of your head, or the fact that the base of your nose is the same width as your eye.

So, both in terms of human proportions and, unsurprisingly, in standard measures, there are three hands to a foot.

This clears up the first part of what we wanted to do; and if hand-sized measurements are important to the job you do or the hobbies you have – if you're a livestock farmer, say, or if you ride horses – that's an important job done.

The Hand, the Space Shuttle and the Horse's Backside

If you've ever seen a picture of a US space shuttle on its launch platform, you may have noticed two white rocket-like appendages, one on either side of the main body. On seeing them, a number of questions may have crossed your mind, like what they're for, exactly, and what happens to them once they're done with, and, if you live under the flight path, what might be the likelihood of one of the things landing on your head, or your house; but of all the questions that might have sprung to mind, the relationship between the size of these rockets and the size of a horse's backside probably wasn't at the top of your list.

But the two are inextricably linked, and I'll tell you why.

The rocket-like appendages are solid rocket boosters, or SRBs, and they work very much like giant fireworks. These SRBs are made by a company called Thiokol at a factory in Utah, and are delivered to the launch site by train. The railway from the factory runs through a tunnel in the mountains. This puts an upper limit on the width of the SRBs: they have to be narrow enough to fit through the tunnel.

The diameter of the tunnel is determined by the width of the trains that run through it, and the width of the trains, in turn, is determined by the width of the track. The width of the track is just over fourteen hands, or the size of two horses' backsides.

The reason that the track is this size is because the USA, like most countries in the world, uses a standard gauge for its railways, known as Stephenson's gauge.

This was the wheel spacing used by the first locomotives, developed by George and Robert Stephenson in the early nineteenth century, and it came about because before the Stephensons there were no such things as trains, and no such things as train-component factories, so the wheels and axles for the first locomotives had to be made by the people who made horse-drawn trams.

These tram builders used the same jigs and tools that they used when building carts and coaches. The same spacing was used everywhere because coach wheels spaced differently would be more likely to get damaged by the old wheel ruts that, over the centuries, had been worn into many of the established long-distance roads by generations of wheeled traffic.

These old wheel ruts were made by generations of wheeled traffic drawn by a maximum of two horses side by side; and the optimum wheel spacing for a vehicle drawn by two horses side by side is roughly the height of a single average-sized horse, or fourteen hands, or the width of two horses' backsides.

And that, ultimately, is what determined the size of the space shuttle's booster rockets.

But we're not finished with our DIY ruler yet. Taking your folded strip of paper, fold it in half. Now fold it in half again, and place your thumb on top of it. What you should find is that the paper and your thumb at its knuckle joint are almost exactly the same width. The paper may be just slightly wider – as wide, in fact, as your thumb with a nail knocked in either side of it. So there are four carpenters' or DIYers' thumb widths in a hand's breadth.

Now unfold your paper and draw a line along each of the folds. Then, starting from one end, number each of the divisions.

And what you have before you now is a foot ruler, created to the scale of your own body and divided, you will find, into twelve parts, which other languages call duims or pouces or pollici – all meaning 'thumbs' – but which in English are known as *inches*, from the Latin *uncia*, meaning 'a twelfth'. And the closer the size of your own feet are to the size-ten standard, the closer your personal units will be to the 'official' ones.

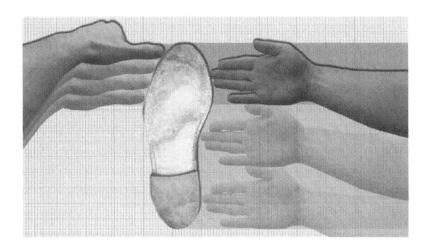

Shoe Sizes, Vegetables and Little Things

I've talked a lot about shoes and boots so far, and how they relate to the width of your hands and the thickness of your thumbs. If you're sick to the back teeth of them by now you can skip this bit and go on to the next section, because I want to talk a bit more about shoe sizes and how they work, and, along the way, about cereal crops and turnips, and how you measure things smaller than your thumb.

You are a shoemaker. Well, maybe you're not, but let's imagine that you are. You've got a big sheet of leather in front of you, and some marking and cutting tools; but because of the way we're doing things in this book, you've got no measuring equipment. Your task is to use your leather to make the insoles for a range of shoes for children and adults.

Let's start with the children's shoes first. If you put your palm flat on the leather and mark out a hand's breadth, that's about the smallest shoe you'll need to make for any child who actually needs proper shoes. You can call that a size zero.

Now mark out a second hand's breadth. That will take you, roughly, to the size of a child's foot just before puberty. So for children, you need to make a range of shoes with insoles between one and two hand's breadths in length.

One more hand's breadth after that will take you to the sole you need for a fully grown man.

But people's feet differ, and you want to make a range of sizes, not just three. The smallest size we've developed in this book so far is a thumb's thickness, or inch; which is all very well for knocking in nails, but too big for the intervals between shoe sizes. A child with growing feet won't thank you at all for a pair of shoes a whole inch bigger than the ones that have just started to pinch.

So what are you going to use for your smaller divisions? Given that there aren't many (or any) obvious body parts that spring readily to mind, and that fit neatly into the system of measures you've developed so far, you've got two choices. One would be to take an inch and halve or quarter it; the other would be to find something common, small and roughly standard-sized that happens to be lying around, and use that.

Both approaches have been used, for shoes as well as for lots of other things; the French, for example, used a quarter-inch scale for shoes for centuries. But for more than centuries, for as long as people have farmed the land and grown the food they eat, they have had seeds and grains, small and handy and evenly sized, lying around in their houses or huts by the sackful; and throughout history these seeds and grains have been used as standards against which to weigh and measure tiny things. And some of these ancient seed and grain measures are still with us today, even though most of us aren't aware of them.

Way back in the days of the ancient Greeks, people used *keration*, or carob seeds, for weighing and measuring, and from them we get the jeweller's *carat*.

In Anglo-Saxon England, the staple crop was not carob but barley. And turnips as well, probably. However, for a number of technical reasons, turnips – even very small ones – never really caught on as recognized sub-divisions of the inch in Anglo-Saxon England. But the barley grain, on the other hand, was used both to weigh with and to measure with.

If you take three grains of barley, from the middle of the ear, and lay them end to end, what you end up with is an inch. And this is how a third of an inch became known as a *barleycorn*. This is a fact you might be interested in or indifferent about, depending upon your outlook; but it would seem to have very little relevance to anything much nowadays, or to anyone save for historically minded barley farmers, were it not for the fact that most of us own at least two items measured in barleycorns, and some of us considerably more; and we use them every day of our lives. This is because one of the few places where the barleycorn has survived, in the English-speaking world, is in the measurement of shoes.

Barley: puts corns on your feet

Here's how it works: the 'base' size of the smallest proper shoe, a child's zero, is a hand, or four inches. The shoe sizes go up in thirds of an inch, or barleycorns, to a children's twelve, at twice the width of your hand, or eight inches long. From there you go on, still in barleycorns, to an adult zero (or child's thirteen) at eight and a third inches and so on up to an adult ten, at eleven and two-thirds inches, and beyond. This is the insole size, remember – so if you add on another bit more – a barleycorn, say – you'll get to your external dimensions, which gives you a size-ten shoe typically a foot long, or three hands.

Unless you live in the USA, that is: there's a one-size difference between the USA and Britain. Our two is their three; and our ten is their eleven. This is because the British call the smallest size a zero and the Americans call it a one. It's the same difference that you find in the numbering of floors in tall buildings: what the British call the 'ground floor', and number as 0 or G on the buttons of what they call 'lifts', the Americans call the 'first' floor and number as 1 on their 'elevators'. More than this, they say 'ee-ther' while we say 'eye-ther', and they say 'nee-ther' where we pronounce the word 'neye-ther'. Contrary to certain views, though, both nations say 'po-tay-toes' and no one but no one ever says 'po-tah-toes'. I hope that makes things clear.

Shoes. There are other ways of measuring them, which don't involve cereals or vegetables. Or elevators or lifts. The most common system outside the English-speaking world is the *Paris point*. It's sort of metric, in that centimetres are involved somewhere along the way, but it's sort of not, as well, because Paris points don't fit neatly into any other metric unit by a factor of ten. Or indeed by a factor of any other whole number, unless you count sixty-six and two-thirds points to the metre as in some way 'whole'.

A millimetre, you see, is too small to be a meaningful difference between one size and the next, but a centimetre is too big, so instead of either of those, a unit of two-thirds of a centimetre was settled on. There were a couple of reasons for this. One is that it is a decent sort of interval, given the speed at which children's feet grow, and given the spread of adult foot sizes. The other reason is that a Paris point is the old French shoe-size unit of a quarter of a pouce, or French inch, by another name. If you want to be really precise, then redefining it as two-thirds of a centimetre does change it slightly, technically, by nought point nought something, but you'd be hard pushed to spot the difference. A Paris point is also, as it happens, more or less halfway between a British half-size (half a barleycorn or a sixth of an inch) and a whole size.

The system, like the English version, starts at a hand's breadth, which is ten centimetres, or fifteen points, in length. This means that the base continental size, the equivalent to an English zero, is a size fifteen; and it goes up from there all the way to the biggest sizes.

And having done feet and shoes pretty much to death now, let's move on up the body a little.

A Stick as Long as Your Leg

A last note on human-scale lengths before we move on to something else.

Sometimes there's a need for a measure a little bigger than a foot to make it easier and quicker to measure out things that come in longer sizes, like string, or carpets, or cloth from a roll.

Here you've got a number of choices. One is to look for a handy-sized body part longer than your foot; and in a number of countries they've settled on the arm, or a part of the arm.

The distance from the elbow to the tips of the fingers has been used in various places at various times. If you take off your shoe and put it against your forearm, you'll find it will fit about one and a half times, depending on whether you measure from the inside or the outside of your elbow. But the fact that you've a choice of how to do it, and that how you do it makes quite a difference, means that the different 'elbow' measures used throughout the world have ended up quite different in size. The ancient Egyptians had two different versions, a regular one of around eighteen inches and a 'Royal' one of almost twenty-one, whilst the Roman version was around seventeen and a half. The Latin word for elbow is *cubitum*, and so this measure tends to be called a *cubit*. It's not much used these days.

For a bigger 'arm' unit, stretch out your arm, and from the tips of your fingers to where your shoulder joins your body is what is known in the eastern Mediterranean and the Near East as a *pik* or *pic*, and the Italians call a *braccio*. In Italy it used to vary by region between twenty-seven and twenty-eight inches, but now is standardized as an informal metric unit of seventy centimetres, or twenty-seven and a half inches.

But the arm, or braccio, has never really caught on around the world in the way that the foot has. The cubit has had more success, but not nearly so much as the foot has. Partly that may be to do with the fact that these units don't fit into such a neat ratio with your other units, whether you measure in feet or metres; and partly it's because if you actually are using your arm to measure cloth you're pulling off a roll, for example, an elbow-length is too short and standing with your fingers outstretched and measuring to the shoulder of that arm probably isn't the best or most comfortable way to go about things. In real life, if you're right-handed, you're probably holding the roll close to your body, tucked under your left arm, and you're holding the loose end in the palm of your right hand as you pull it out. And when you reach the furthest stretch, that's the length you've measured out. It's a bit longer than a braccio; but, as it happens, it is a distance that seems to have a lot going for it as a unit of measure.

From the palm of one outstretched arm to the opposite shoulder of an average person is, pretty much exactly, three times as long as the sole of your shoe. And also, as it happens, as long as your leg, from your heel to your hip; which means that if you stand up straight, and imagine holding in your hand a walking stick of just comfortable length, that's how long it would be. Now it just so happens that for thousands of years people around the world have been carrying around 'instruments' of roughly this size, which they have used to walk with, and to herd their cattle with and – occasionally – to measure with. As a result of this, two of the most widely used measures on the planet – the *yard* and the *metre* – are both sticks about as long as your leg. The word yard, by the way, is derived from an old word, *gyrd*, which means 'stick'. So a yardstick is, actually, a stickstick.

A stick

I'm sorry. At this point I need to break off, briefly, to deal with a couple of irritating and insistent voices which I keep hearing in my head, and which just won't shut up. One is telling me that, actually, the metre has nothing at all to do with sticks, or legs, or anything else to do with everyday life. What the metre was designed to be, this voice is telling me, is exactly one ten-millionth of a *quadrant*, which is the distance between the equator and the North Pole, as measured on the Paris meridian, assuming that the earth is a perfect sphere – which it isn't, as it happens. But if it were, that's what the metre would be. The other voice, meanwhile, is telling me an altogether more noble story about the yard, and how the Norman King Henry I decreed, back in the twelfth century, that the yard should henceforth be the distance from the end of his royal nose to the tip of his outstretched finger.

A plague on both your houses, I say. And may I respectfully direct your attention to something known as the *Intentional Fallacy*.

The Intentional Fallacy – or why you shouldn't always believe what the 'experts' tell you

The 'intentional fallacy' is what happens when people fail to realize that what things actually are, for all practical intents and purposes, isn't always the same thing as what their owners or creators intended them to be. I'll give you an example: the works of William McGonagall, the excruciatingly bad Scottish poet. Now, it's likely that Mr McGonagall wrote his poems intending them to be read seriously, and perhaps even to bring a lump to the throat and a tear to the eye. Here are the closing lines of one of his most famous works, 'The Tay Bridge Disaster':

> *Oh! ill-fated Bridge of the Silv'ry Tay,*
> *I must now conclude my lay*
> *By telling the world fearlessly without the least dismay,*
> *That your central girders would not have given way,*
> *At least many sensible men do say,*

Had they been supported on each side with buttresses,
At least many sensible men confesses,
For the stronger we our houses do build,
The less chance we have of being killed.

And when McGonagall gave public readings, people laughed. Not only did they laugh, but they came from miles away just to do so. Some of the readings became so crowded, and so raucous, that they ended in riots.

So whatever these poems were intended or designed to be matters very little: what they actually are, in the real world, are comic works.

Getting back to measurement: to any vaguely normal person a metre has nothing at all to do with ten-millionths of quadrants or whatever the current 'official' definition is – in use, it's actually a stick or ruler about as long as your leg. Ditto the yard: and it's no use appealing to Henry I – he's long since dead, and can't help you. It is these units' quality of stick-as-long-as-your-leg-ness that has made them successful and accepted around the world while other measures with equally august or scientific-sounding origins have withered and died.

While I'm on the subject of origins and history I'd like to take a brief detour to revisit the world of Alexander Thom, the Oxford professor who, you may remember, took measurements of large numbers of Neolithic monuments and worked out that they all seemed to be built using a common system of measurement. The most important measurement for Thom was what he called the Megalithic yard – a measure which at 2.72 feet or 0.8297 metres, was some three inches short of a modern yard. This difference in size is very important for some of Thom's followers today. They point out the fact that 366 Megalithic yards is a distance equal to one second of the arc of the earth's circumference, and take this as evidence that the people of the late Stone Age were possessed of advanced mathematics and geometry.

Which may or may not be so. But equally – and given the fact that the average height of a Neolithic man, like his yard, was about three inches shorter than the modern version – it's not beyond the bounds of possibility that the people of the late Stone Age were possessed of sticks.

One last measurement I'd like to briefly touch on is the measure you get when you stretch both arms as wide as they will go, with your fingers open. It's not a great deal of use for market traders or others who measure out cloth or string from rolls, because you can't do a great deal with your hands while you're standing like that – not even grip your cloth. And besides, in a crowded marketplace there's always the danger of poking someone in the eye.

But it is a very handy measure indeed for people who do things with boats and water. This is not so much because of the arms-outstretched nature of it, but because, at six feet, it's the depth at which everyone of average height is in over their heads, and at which even most taller people will find the water level higher than their noses. It's the 'out-of-your-depth' measure, and the measure which enables you to visualize how many times deeper the water is than you are tall. It's called the *fathom*, in English, which comes originally from the ancient proto-Germanic word *fathmaz*, meaning 'an embrace' and goes by various other names in other countries, including the *favn* in Denmark, the *famn* in Sweden and the *ken* in Japan.

Measuring the world by 'smoot'

The fathom has a much younger and slightly shorter cousin. It goes by the name of *smoot*. If a fathom is the width of a man's outstretched arms, then a smoot is the height of a man. Not a man in the sense of men generally, or even an average man, but one man in particular: Oliver R. Smoot, the five-foot-seven-inch Chairman of the American National Standards Institute and President of the International Organization for Standardization.

In 1958 Mr Smoot was a student at the Massachusetts Institute of Technology. He was also a 'pledge', or novice member, of the Lambda Chi Alpha fraternity. As part of their initiation, a group of pledges was given the task of measuring out the length of the Harvard Bridge, and they were told to use one of their number as a 'ruler'. Smoot was chosen for the job because he was the shortest – which would make the task longer and more arduous – and because he had the silliest name.

The measurement was done at the dead of night by getting Smoot to lie down, and marking his height with chalk and paint. He would then stand up once more, move one length further along and get down again, and again and again. For a while, Smoot did it under his own steam but after a hundred or so times he became tired, and his companions ended up simply dragging him from one space to the next.

Measuring by smoot

In this way, it was determined that the bridge was 364.4 smoots long 'plus epsilon', although this was later recorded as 364.4 smoots 'plus an ear'.

That would have been the end of it, but for the fact that the strange new markings on the bridge caught people's fancy, and it became the custom every year for each new intake of Lambda Chi Alpha pledges to repaint the markings.

This went on for some time, but in 1987 the Massachusetts Department of Public Works decided that the bridge needed renovations and resurfacing, and this meant removing all the smoot markings. This caused something of a commotion locally, and the press contacted Oliver Smoot, who by then was forty-eight years old and executive vice president of the Computer and Business Equipment Manufacturers Association in Washington DC. Smoot was asked whether he would be prepared to be re-used for new markings, should the need arise. He was less than sure that he would.

Meanwhile, the Massachusetts Metropolitan District Commission, the government body in charge of the bridge, went on record in support of smoots. 'We recognize the smoot's role in local history,' they said. 'That's not to mean that the agency encourages graffiti painting. But smoots aren't just any kind of graffiti. They're smoots! If commemorative plaques and markers are not installed by the state once the bridge work is done, then we'll see that it's done.' The Boston police authority also weighed in, saying that its officers had come to rely on smoot-marks for identifying the location of accidents on the bridge. Faced with the strength of opinion, the Department of Public Works obliged; and the contractors, Continental Construction Company of Cambridge, also agreed to make the new concrete paving slabs five feet seven inches long – instead of the usual six feet – to coincide with the smoots. On completion of the work, Lambda Chi Alpha were given permission to resume their annual smoot-painting ceremony, which continues to this day.

For a long time, smoots were what you might call a local measurement, much loved where they were but pretty much unknown to anyone outside. However, because of the kind of people that MIT graduates are, and because of their involvement in things like the development of the Internet, smoots these days turn up in all sorts of odd places. You can even, for example, select smoots as your chosen unit of measurement in the Google Earth software.

In recent years there have been one or two attempts to 'modernize' the smoot by bringing it into line with the height of Smoot's son Steve (MIT '89) or his daughter Sherry (MIT '99) but so far all such initiatives have been rebuffed by purists.

Pick Your Own Numbers

So far, we've arrived at some units for measuring and calculating human-scale distances, and related them to a set of simple numbers: four 'thumbs' or inches in a hand and twelve in a foot, and three feet in a stick or yard as long as your leg.

In getting to where we are, and in sorting out the numbers, we haven't really had a great deal of choice. The numbers we have used have been the numbers that actually exist in the world in the ratios between the sizes of real feet, real hands and real thumbs.

But not all measures are like that, and at some point very soon we're going to have to start making decisions about what sorts of numbers we want to use in our measuring. We could decide to divide everything by ten, say, or twelve, or three point one four two, or sixteen, or three hundred and sixty, or any other number or set of numbers we like. The question is, which number or numbers work best, and for what purposes?

And here is a good place to do another little experiment, to find out.

You will remember the way we worked out how to divide up a foot, by folding a strip of paper. Well, if you have any of that paper left, now is the time to dig it out. You will need a fair amount of space for this – something like a kitchen table would be ideal.

First cut a dozen or so narrow strips of paper. It doesn't really matter how narrow they are, or even how long, except that they should all be the same length, and long enough to divide up into ten or twelve sub-units.

Now, take a strip of paper and, picking any number you fancy between two and twelve, cut your strip into that number of equal-sized pieces, going by eye alone, and without using a ruler or measuring tool of any kind (not even the paper 'ruler' you made earlier).

When you've done that, lay the pieces side by side and see how accurate you've been, and how close they are to each other in size. And be aware of how easy or difficult it was to do. Then take another strip of paper and do the same to that one, but choose a different number of pieces to cut it into, and so on and so forth until you've used up all of your strips.

By the end, your table-top will be covered in different-sized groups of cut-down paper strips. For some groups it will have been quick and easy to cut accurately, and for some groups it will have been harder and more time consuming. Make a note of the one you found easiest to do, and also of the one you found second easiest.

Now count up all of the pieces on your table and write down the total.

Here's what I found when I did the exercise myself: I found it easiest and by far the most accurate to divide by two. The next best, I found, was to divide by three. And for cutting up bigger numbers I found it better to do it in a number of smaller steps: to cut four pieces it worked best to cut in half and half again, rather than to judge individual quarters; and to cut in twelve it was almost impossible to judge an accurate portion by eye, but far easier to cut into three and then two and two. The trickiest ones to cut were the bigger prime numbers you couldn't break down into smaller chunks of two or three – five, seven and eleven – and their multiples, numbers like ten.

But the other thing I found – and you will have found, too, unless you are very strange indeed – is that when I added up all my pieces at the end I wrote the answer down as a number in *base ten* – which is to say, this many tens and that many units. And I'll be very surprised indeed if you, or anyone else, so much as even considers doing it in any other way.

So the lesson of the exercise is that of all the numbers you could possibly use, there are just three that really work. These numbers are two and three for doing the dividing, and ten for the counting.

Three Ways to Cut Your Cake

The message behind all that cutting-up of paper we've just done is this: you only need three numbers to measure with. These numbers are two, three and ten.

But it might seem, looking at all the different systems and units out there, that you actually need rather a lot more. There are weights that divide into thousands, and angles into fractions of three hundred and sixty, and we've already made ourself a ruler that divides into twelve. But if you look a little more closely, you will see that almost every single number used by every single measure that actually works, and gets used, is the product of either just one, two or, occasionally, all three of the three basic numbers.

Take the number two. What you end up with, when you cut a thing in two over and over again, is a series of numbers that scientists and mathematicians call the *binary sequence*. The first numbers in the sequence are 2, 4, 8 and

Divide by	2	2	2	2
Share	½	¼	⅛	¹⁄₁₆

16; and that is about as far as you generally need to go for most everyday purposes – cutting cakes, for example. Even if you've started with a fairly large thing, by the time you've got it down to a sixteenth of its original size it's going to be a lot more manageable.

However, computer scientists, who do a lot of dividing and multiplying by 2, go much further, with 32, 64, 128, 256 and 512 being common values for memory devices, amongst other things.

Divide by	3	2	2	2
Share	⅓	⅙	1/12	1/24

But sometimes you can't get everything you want simply by dividing things into two. And the best and simplest way to vastly increase the number of ways in which you can divide something up is to add in the number three somewhere along the way.

You only need to divide by three once, and then you can go back to your simple half-and-half-again divisions, but every number you arrive at from then on will be capable of being divided by multiples of both two and three.

The official name for this sequence is *duodecimal,* but I prefer to think of it as 'modified binary' – a sequence of twos with a three thrown in. This use of threes is very common and particularly so for things that are judged by eye.

There is a reason for this, and it's one that artists and photographers have known about, or grasped by instinct, for a very long time but which most of the rest of us can go through our whole lives without ever being consciously aware of. Although dividing by two is the simplest of all divisions, and although it is perhaps the easiest thing of all to compare two things rather than any other number, the human eye, coupled with the human brain, has an inbuilt tendency to divide up what it sees into three roughly equal parts; and we find it natural and easy to see things in thirds. At this point you may well be thinking that your eyes do nothing of the kind, nor your brain, neither; but bear with me for a little while and I'll explain. As a bonus, you may even learn how to take better photographs, too.

The Rule of Thirds

You're reading this now, and looking at these words, but unless you have very unusual eyes, these words – or whatever you happen to be 'looking at' and concentrating on right now, are only a fraction of all the things you could see, without moving your eyes away from the page, if you stopped concentrating on what you're reading. Try it for a moment: keep your eyes as still as you are able, and keep them looking here, but be aware as much as you can of what's going on elsewhere, on either side, and above and below.

Let's work this through, and let's start with what's happening on either side of this page for the moment.

A first observation: there are more things your eyes can detect, from left to right, than are contained in the words on this page. But if you're not 'looking at' them, and if you're not bored with this book yet, you don't really pay them much heed while you're reading.

Now, the size of the bit in the middle that you're looking at – your field of central vision – will vary in size, depending on what you're doing. When it comes to reading, like now, it can be quite narrow. For other purposes, such as walking down the street or watching a film at the cinema, it will be wider. But in the normal course of things the bit you are paying the most attention to is almost always within the central one-third of what your eyes can see.

The way human eyes are arranged (on the front of our faces, generally), allows us to 'see' things to a greater or lesser degree in a sweep of just over a hundred and eighty degrees in front of us – imagine swinging your arms out from shoulder to shoulder and that's about it. Some of that will be real corner-of-the-eyes stuff, hardly seen at all. About a hundred and twenty degrees – the area bounded by the lines marked as 'a' on the diagram – we are able to see properly, with both eyes at once. And of that, we typically give our full attention to just the middle bit, a maximum of sixty degrees – the central part of the diagram bounded by the lines marked 'b'. This accounts for one-third of our total field of vision.

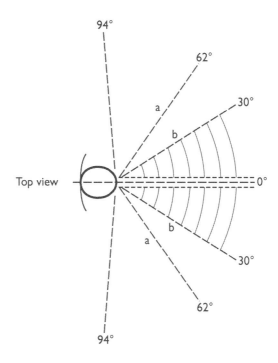

So far it's a bit of a 'so what?' though – the fact that we 'look at' the things we happen to be pointing our eyes towards probably isn't news to you. But there's a bit more to it than that. There's something else. And do keep with me for just a little longer – this really does all come back to the use of the number three, even if it seems like we're going way off piste right now.

As I said, there's something else; and that something else is to do with the fact that our eyes are rarely ever still. They see things, take in what there is to be taken in and then move on to the next thing. As you do when you read this book, in fact. So what we happen to be looking at right now is sort of interesting to us, as far as it goes, but it quickly becomes old hat – and by 'quickly' I mean in a matter of less than a second. What's even more interesting to our eyes than where they happen to be looking at any given moment is where they're going to be looking next. And however interesting the thing we're looking at, there always is a 'next': the next word on the page, the next face in the crowd. Something just outside our main field of attention gets spotted, and gets earmarked 'potentially interesting', and our gaze is drawn towards it. And our interest in it reaches its peak just at the point where it crosses from being 'peripheral' to being in the full headlight-beam of our gaze – at the boundary of the central one-third of the field of vision. After that, when we've checked it out properly, we're quickly done with it and ready to move on to the next thing.

And this brings us to artists, photographers and a thing called the *rule of thirds*, and to how to take better photographs; and, ultimately, to measurements.

The best way to understand it is to do a simple practical exercise, using a print of a famous picture or painting – any famous picture or painting of your choice – and a ruler (just this once!) and pencil.

All you need to do is to take your picture or painting, measure a third of the way in from the left-hand side and draw a line down from top to bottom. Then do the same

a third of the way in from the right-hand side. Now, measure a third of the way down from the top and draw a line across from left to right, and another one across, a third of the way from the bottom, so that you end up with your picture divided into what looks like a large noughts-and-crosses board of nine squares; and then see what you see.

What you should see, in pretty much every case, is that in 'good' pictures which large numbers of people find appealing or moving, the most important elements of the image almost always sit on or close to the lines that mark the horizontal or vertical thirds, or at the intersections between the two sets of lines.

Where this is all leading is to the fact that the number three is important to the way that we see the world; and because of this, measurements that involve judgement by eye will often involve at least one division by three.

So, two hands that find division by two the easiest to do; two eyes that can do twos just fine, but find a sort of natural symmetry or attraction in the number three; and then there is the thinking mind, which works to a different set of needs altogether; and which gets on very well indeed with the number ten.

Divide by	10	10	10	10
Share	$\frac{1}{10}$	$\frac{1}{100}$	$\frac{1}{1,000}$	$\frac{1}{10,000}$

We have ten fingers to count with (and some of us still use them that way from time to time!); we have a system of numbers built around base ten; it's the easiest number by far to use for mental arithmetic. When we think of numbers, most of us think, most of the time, in tens.

It's not always the best number for simple physical division – like cutting up cakes, for example. Using tens, or parts of tens, it's hard to get to the portions that most people need to divide things into. You can do a half, but not a quarter, and not a third. You can sort of get round this by using hundreds – a quarter of which is twenty-five – or even thousands – a third of which is three hundred and thirty-three, as near as dammit; but then this brings its own problems. Biggest of these problems is the huge gaps it leaves in your range of handy-sized units when one unit is too tiny for the job in hand and the next one, a thousand times larger, is just too big.

It's interesting to note that when people with 'thinking' jobs and 'doing' jobs work together on the same project – architects and builders working on the construction of a house, say – the architects, who do the large-scale thinking and planning, often work in scientific base-ten units which the builders, who are more concerned with immediate practicalities, will then translate into their more practical and human-proportion-based base-two and -three measures with their size-ten measuring boots and their thumb-width nail spacing.

Other portion sizes

There are sometimes – very occasionally – situations where you come up against a task that leaves you feeling the need for a number outside the extensive range covered by the three basic number sequences.

More often than not, though, the best solution is not to look elsewhere – it is simply to recombine the same three 'do-it-all' numbers, but just in a slightly different way.

Twenty (2×10), for example, has sometimes been used to divide; in Britain's pre-decimal currency there were twenty shillings to the pound, and in Britain and America today there are twenty hundredweight to the ton. It's not quite as good for mental calculation as ten, and not quite as easy to use in practice as base two, but some people, at some times, have found it a useful compromise.

And different combinations of all three numbers, like sixty $(2 \times 3 \times 10)$ and three hundred and sixty $(2^2 \times 3^2 \times 10)$ are used for time and for the measurement of angles, in which it is important to be able to divide things up in as many different ways as possible.

This is not to say that there aren't, or haven't been, systems of measurement based on numbers other than two, three and ten.

The number seven is one that pops up from time to time, seven days to the week being the most common example. You can sort of see how it got that way: if you live in a farming society where the stages of the moon are important, someone or other will have worked out that the moon takes around twenty-nine and a half days to complete its circuit. If you do a basic binary division on that, half and half again, you end up with seven (or seven and a bit, which is much the same thing) in whole days. So a week of seven days is a useful rough-and-ready bodge to divide a month into quarters. But actually, as a number to use for other things, it's not particularly easy or natural. But it has had a certain sort of appeal over the ages, particularly for those with mystical inclinations – if you spend any time looking up at the sky, as mystical people tend to do, and if you don't have a telescope, you might be able to spot the seven celestial bodies in the solar system visible to the naked eye – the sun, the moon, Mars, Mercury, Jupiter, Venus and Saturn – and you might come to the conclusion that seven is all there are. Couple that with seven colours in the rainbow, and it begins to look like there's some sort of higher purpose behind it. The result of that is that religion and mythology tend to be packed full of sevens. There are the seven deadly sins of Christianity, the seven-branched menorah of Judaism, the seven heavens of Islam and the lucky seventh son of a seventh son of Irish folklore, to name but a few.

But when it comes to measuring, as numbers go it's not that easy to use, and you'll probably find it simpler to steer clear of sevens and other awkward numbers if you can manage it.

The Human Weighing Scales

So far we've dealt with measures that don't give us much choice. Your feet and your hands and arms and legs are there to measure with, and there's a natural ratio between their various sizes. So it's hardly surprising that people around the world have independently come up with similar-sized and similarly divided units for human-scale lengths.

But when it comes to weighing, you might think we're moving into a different sort of area.

Most of us don't have body parts we can take off and plonk down into a scale pan, and so we need to look to the outside world for our standards of measurement.

When we come to look at the world around us for things we might use as a standard of weight, the things we could conceivably choose come in any number of shapes and sizes, ranging from specks of dust through grapes, oranges, staplers and wellington boots to boulders and elephants. Tropical countries have coconuts, while in snowy ones it's possible to get hold of snowballs as small as your fingertip or as big as an ice-fishing hole or even an igloo.

So you might think that when it comes to setting down standard measures of weight, different times and different places would come up with very different units.

But when you actually come to look at some of the standards of weight that people have used over the past few thousand years, what is surprising is just how similar they are.

To find out why this should be, let's imagine that you are creating a standard unit of weight, right now, to be used both in the home and out and about, for everyday purposes.

The first thing you will need to decide on is how you're actually going to do the weighing, and also how the people who use your unit are going to use it. Now, it is possible that you may have the right sort of professionally made measuring equipment on you right now, or, if you haven't, you may be willing and able to spend a bit of time improvising a set of scales out of materials you have to hand; but you can't assume that others will, or even that you will, on every single occasion that you need to estimate the weight of something.

So you're going to be judging the weight yourself, by feel.

This means that you will need to be able to pick up whatever object you've got in mind as your standard weight. This starts to rule things out and narrow your range a little: so, no cars, no elephants, no tree-trunks, no concrete slabs.

The next thing to be aware of is that to weigh other things against your new pickupable standard weight, you're going to have to be able to pick those things up as well, at the same time as your standard weight. To compare the weights, and feel the difference between the two, you will need to be able to hold each, at the same time, in different hands.

This rules out small boulders, or large pieces of machinery, or anything else that takes two hands to lift. But it does – for the moment – rule in anything light enough to pick up with one hand.

However, some of the things you can pick up with one hand can be too light. If it's not weighty enough to feel the effect on the muscles of your hand and arm, it will be hard to tell the difference when whatever you're weighing against it is lighter. So it is probably for the best, for your main unit, to rule out things like grains of sand and small pebbles.

So, from an almost endless list of possibilities we've come down to a fairly narrow range. And there are considerations that might make you want to narrow it still further: if you're going to use your weight regularly, for example, or if you're going to keep hold of it for any length of time, it's probably for the best not to pick something right at the upper limit of what you can hold in your hand, or something which, if you dropped it on your foot – which does happen – would smash your toes, even if you were wearing shoes.

And finally, if it's a weight for everyday purposes, you will probably want it to relate to the everyday sorts of things you might need to weigh out, such as ingredients for a family meal.

To get a sense of how narrow or broad a range we're left with, it's worth taking a look at the units that different societies have used over the years, and how they compare in size.

What you find is that almost every society, both now and in the past, has, or has had, a unit of weight that, if it were a smooth, rounded stone, would nestle comfortably but weightily into the palm of one hand, so that your fingers would curve around it but not fully enclose it.

From a potential range of zero to infinity, the lower limit for these stones-that-fit-comfortably-into-the-palm-of-your-hand weights seems to be around the size of a Roman *libra pondo*, or *pound,* and the upper limit – with one

KILOGRAMME

CHINESE JIN

MALAYSIAN CATTY

THAI CATTY

JAPANESE KIN

HEBREW MA'NEH

VIENNESE PFUND

METRIC JIN

METRIC LIVRE (informal)

DUTCH POND

LIVRE DE PARIS

LIBRA (Portugal)

LIBRA (Spain)

LIBRA (Italy, varies)

PFUND (varies)

POUND (US/UK)

STOCKHOLM PUND

RUSSIAN FUNT

TROY POUND

ROMAN LIBRA

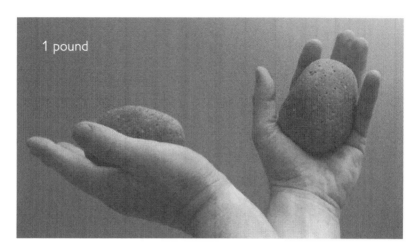

A stone

exception – at about the level of the *jins* and *cattys* of the Far East.

The one exception, standing out on its own at almost a pound heavier than its nearest 'rival' is the *kilogramme*.

As with the metre, it was never the intention of the people who invented the kilogramme that their unit should be judged on anything as imprecise and unscientific as the ability to fit neatly into your hand; but for the practical, everyday purposes we have in mind here it does have quite a few of the qualities of the other hold-in-your-hand-sized units. If you find a stone that weighs a kilogramme, you can pick it up without too much trouble and you can hold it in your palm. It's just that no one else seems ever to have picked something this big for their

1 kilogramme

Another stone

main unit; or, if they did, it seems not to have caught on for long enough to have been recorded. More than this, almost all countries that have had the kilogramme for any length of time have developed a smaller, unofficial 'metric pound'. Mostly it is a *livre, Pfund* or *jin* of five hundred grammes, although in Thailand and much of South-East Asia they prefer a 'metric' *catty* or *kati* of six hundred grammes (which, as it happens, is only four and a bit grammes lighter than the pound) and a third 'non-metric' catty still used in Malaysia. Whatever qualities the kilogramme has as a unit of weight, it seems that around the world it doesn't fully meet whatever requirements ordinary people have for an everyday unit of weight.

One likely reason seems to be that when you pick up

things which are less dense than stones – fruit and vegetables, say, or a lot of the other kinds of food items you typically measure out by weight – a pound is about as much as you can comfortably hold in one hand. For the purposes of research I did actually manage – after a fair amount of effort – to get a whole kilogramme of apples balanced in my hand (see picture). I managed to keep them there for about thirty seconds, but the little pyramid I constructed collapsed before I could get a photograph and I dropped them all over the floor.

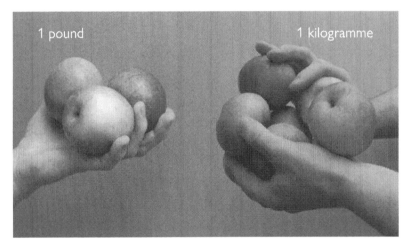

Not stones, but apples

For what it's worth, I have another little story of my own to add about weights. It may or may not be relevant, but it concerns the kilogramme-sized stone in the picture on page 85. My daughter Alice, when she was very young, was playing by the edge of a stream in Scotland, and collecting smooth round stones from the water's edge. She already had a small armful, but wanted one more; and as she crouched down to pick it up the stone from the top of her pile rolled off and dropped the short distance – probably no more than three or four hands' breadths – onto her finger. It smashed the bone, took off the nail and partially amputated the fingertip itself, which was hanging on by a thread of skin. Luckily, the doctors managed to sew it back on again. I keep asking her and her sister Lucy to try it again, for the sake of science, with a range of stones from a Roman *libra pondo* upwards, but for some reason they always refuse.

The last thing I'll say about the kilogramme before we move on is that it's rather a long name to use in everyday speech, so it tends to get shortened. The most common shortening is to call it the *kilo*, which is pronounced *kee-lo*, because that's how you pronounce a word spelt like that in French. I could be completely on my own here, but when I hear people in Britain pronouncing it French-style it always sounds like a bit of an affectation to me. To me, it sounds like when people go on holiday and then they tell you about where they've been, using ostentatiously 'authentic' pronunciation. And I have a particular thing about the ones who do the 'sound effects' too: it just bugs me. But maybe, as I said, it's just me.

Pebbles and gravel

Once you've got your unit of weight, you'll need to decide how to divide it up.

You do have a number of options here, and there have been different ways of doing it over the years, but the way your body is set up, and the way most scales and balances work, tends to lead most people down one particular route.

You have two hands to weigh *this* against *that*. The most common forms of scales and balances have two pans, one on *this* side and the other on *that* side, to compare one thing with another. It's far less common, and far more difficult, to compare more than two things at the same time, and this would be true even if you had three hands, or even if you invented a set of scales with three or four or five pans.

So, in general, measures of human-scale weight have tended to be divided in twos. One of these makes two of those. Two of those makes four of the next size down. And so on, half and half again until you get down in size from a weighty palm-sized stone to a smallish but still feelable-in-your-hand pebble one-sixteenth of the size.

You find this division of weight all around the world. The Japanese split their national weight, the kin, into sixteen *tael*; the Malaysians have sixteen *tahil* to the catty; the British and Americans sixteen *ounces* to the pound. Until the advent of the metric system – which we'll come to in a moment – the Chinese traditionally did likewise, dividing their *jin* into sixteen *liang*.

Other numbers in the binary series besides sixteen have been used as well, from time to time: the Germans divided their *Pfund* not just into sixteen *Unze*, but further, into thirty-two *Lot*, whilst the French used a half-livre *marc* as well as sixteen *onces*; but of all the ways to divide a weight, half-and-half-again binary is the most common method by far, and sixteen the most common number of the lot.

However, a smaller group of countries have done it differently: mostly these countries have gone for a twelve-ounce pound. The old Roman pound was split into twelve *unciae* – the word itself, from which 'ounce' comes (and 'inch', as we have seen), means 'a twelfth' – and the regions that spent the longest under Roman occupation – namely much of Italy and southern France – seem to have stuck with doing it that way for longest. And also, and probably because of the divisibility by three and the greater range of dividing numbers that fit into the number twelve, for greater precision the twelve-ounce *Troy pound* used in the French market town of Troyes was for a long time widely used by jewellers all around the world.

The metric system is a different kettle of fish altogether. Compared with other systems it has unconventional sizing; it has unconventional division; it goes straight from an unusually large rock-sized kilogramme down to a tiny gravel-sized *gramme* with nothing in between. Its units are based on things you find in physics textbooks, rather than on the kinds of objects that normal people deal with in their daily lives.

But it does have logic and rigour and intellectual simplicity going for it, and – at least in its 'official' form – it has a stern unwillingness to adapt itself to the messy and imprecise bodging that goes on in everyday life. All of which makes it extremely precise and constant, and extremely good for scientific calculation. A British pound may be bigger than a Roman *libra pondo*, and smaller than a Japanese *kin*, but a kilogramme's a kilogramme wherever it is.

In the laboratory, the kilogramme is hard to beat; but in the kitchen, and in the marketplace, people tend to customize it and 'fill in the gaps' to make it a better fit for the more mundane purposes they have in mind.

The main way in which it is customized, as we've already seen, is by dividing it in half, or a bit more than half, to come up with a unit that approximates to a pound or to a catty. But each of these units still goes straight from quite big to exceedingly tiny, and misses out the pebble-sized ounce or tael unit.

There are two ways that people deal with this. One is to think of it as a *modular* unit of various handy-sized sub-units – but without committing the heresy of giving the sub-units their own names. The other is to go the whole way, and to turn the kilogramme into something much closer to the old units it was supposed to have replaced.

Outside the laboratory, the modular way of dividing the *kilogramme* is more common than any other way. No one at all ever goes into a shop and asks for three hundred and forty-seven grammes of potatoes, or six hundred and twelve grammes of grapes: what they ask for are multiples of a set of different 'sub-units' that go something like the diagram overleaf.

Which is to say, people tend to think of the kilogramme, in daily life, not as something with a thousand equally important divisions, but as something divided into a much smaller number of different-sized 'chunks'. These chunks fit one into the other in a more-or-less binary way, for the most part, with a couple of almost-binary two-and-a-halfs and a three thrown in to make the numbers fit.

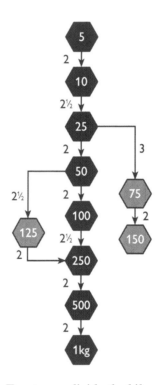

How Europeans divide the kilogramme

Of all these 'chunks', the one that seems most common and fundamental to most of the different sizes that people use, and that goods are sold in, in metric countries, is a chunk of twenty-five grammes. So you will often see a seventy-five gramme pack, or a hundred and twenty-five gramme pack, but you won't so often see a ninety-five gramme one, or a hundred and five or hundred and ten gramme one.

Twenty-five grammes feels roughly right for two reasons. One is that it's a useful small-but-noticeable size for everyday purposes; the other is that twenty-five is easy-ish to multiply in your head. And as it happens, twenty-five grammes is almost exactly the same size as the basic 'chunk' used in traditional systems, the *ounce*, which weighs in at around twenty-eight grammes.

There's still a little bit of mental arithmetic to do, though, as there always is when big numbers are broken down into chunks. In counting out quantities, you always have to remember to multiply the number of chunks you're counting by the value of the chunk itself – just like doing your times tables at school. So fifty grammes is two chunks, which is one chunk – or twenty-five grammes – smaller than seventy-five grammes. It sounds complicated when you write it out like this, but people who grow up with the system eventually learn to do the translation automatically in their heads. It's a bit like tennis scores: you just come to know that one point is fifteen, and the next point after that is thirty, not sixteen.

In China, they have gone a bit further in dividing up the kilogramme to make it convenient to use, even for people who don't know their times tables. They've either gone such a little bit further that you'd hardly notice the difference, or else they've gone so much further that it's completely unacceptable, depending on your point of view.

What they've done is to formalize a couple of the natural subdivisions of the kilogramme, and given them names and official status as units in their own right. They've named five hundred grammes – just over a pound – a *gongjin*, and fifty grammes – just under two ounces – a *liang*, which makes the official set of units shown in the diagram.

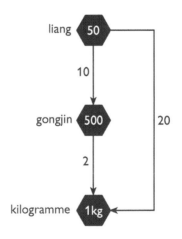

How Chinese divide the kilogramme

From a practical, normal person's point of view that's no big deal: all that's happened is that a couple of the chunks that people use anyway have just been given names. From a normal person's point of view it's just a formalization of what's already happening, and a small shift from *de facto* to *de jure*. Even if words like *formalization* – and indeed *de facto* and *de jure* – aren't quite the words that a lot of people would use to describe it. If anything, the change has simplified everyday measurement, and made it easier to work with in your head. One chunk of fifty is just 'one', not fifty; three chunks are 'three', not a hundred and fifty; and so on and so forth. This also has the effect of making the system more decimal and 'metric' in everyday use rather than less so, by dividing the most commonly used quantity, the gongjin, into ten, rather than the mish-mash of binary and almost-binary subdivisions that are common in other countries.

But from a theoretical point of view, this thing actually poses quite a big problem for the metric system. From a theoretical point of view, it means accepting that everyday convenience is sometimes as good a reason for a unit as scientific purity. And it also means accepting that sometimes some other number than a ten, a hundred or a thousand might be the best for the job.

Once you accept two gongjins to the kilogramme, for example, or fifty grammes to the liang, then where do you draw the line? What is it that makes a twenty-liang kilo all right, and a twelve-inch foot not all right? And where does that leave the consistency of the metric system?

But on the other hand, if your answer is just to ban all these gongjins and liangs, and all the other not-quite-pure metric units, and if you studiously and pointedly ignore anything that might improve the everyday convenience of the people who use your system, then the question is, why should people want to use it at all?

These are not easy questions, and we'll come back to them later on.

The Long Walk

Let's take a walk.

In fact, let's make it quite a long walk. What with all these liangs and gongjins, and what they mean for the integrity of the metric system, it all gets a bit much. You could probably do with a bit of fresh air.

But there's just one little thing . . .

I want you to tell me, when you get back, just how far you've walked.

And remember: the usual conditions apply here – no copying, no cheating, no using anyone else's work. So no pedometers, no rulers and maps, no mileposts, no anything that uses any systems or equipment already invented by anyone else. Think of this as the Year Zero as far as the measurement of distance is concerned. Someone's going to have to invent it all from scratch, and that someone is you.

So how are you going to do that?

You could do what you did for your room – walk it

out in little steps, one foot directly in front of the other; but it would get a bit tedious after a while, as well as slowing you down no end and generally getting in the way of the pleasure or the purpose of your walk. And you'd look silly, too.

The other thing you could do would be to walk normally, but count your steps as you go; and then you can pick a suitable number of paces to use as your unit and leave it at that, or else do a bit more work to relate it to the size of your feet, or hands, or walking stick or whatever.

Now, if you're going to walk and count there are a couple of ways you can do it. You can count every time you put a foot down – left, right, left, right, one, two, three, four – or else you can start with your leading foot and count every time you put that down – left . . . left . . . left . . . one . . . two . . . three . . . , and so on.

It all comes down to personal choice, but most people find the single-step quick-counting a bit irritating after a while. If you look at soldiers, who spend a lot of time counting while they walk, they tend to go for the longer-spaced double-pace counting for long marches.

So, double paces, then. And because the counting is going to be done in your head you're going to be doing it in base ten. Base ten is also good here because you have ten fingers, and if you're worried about losing count in your head, you can always tuck one finger in every time you've reached ten paces, and the next one when you reach twenty, and so on up to a hundred. If you're going to do this a lot you can even improvise a makeshift tool out of knotted string to hold between your finger and thumb to count off your tens or hundreds as you go.

By the time you've gone a hundred double paces you've made a decent start; and by the time you've gone ten times that you can probably afford to give yourself a little pat on the back.

Because you're counting in tens, hundreds and thousands, and because you've covered a reasonable distance by the time you reach a thousand, here is probably a good time to set your unit of distance, and start counting from one again. And anyway, you won't want to be counting numbers in your head that take longer to say than the time it takes to make the step (one thousand seven hundred and seventy-seven, one thousand seven hundred and seventy-eight . . .).

You can give your unit any name you like: if you speak English you might choose something simple and descriptive like a 'Thousand-Pace-Unit' or 'Thou'. Or else if you speak Latin, you might want to call it a *'mille passus'* or *'mile'*.

Now, once you've got your unit, you've then got two things to sort out. One is to work out how big it actually is, in relation to your other units of measure, and the other is how to divide it up. And you might even add a third thing to sort out, which is, which of these two issues to address first?

The most logical way to go about it would be to do the first part first, which is to say it would make sense to know how long a mile is before you set about marking off fractions of a mile. But funnily enough, what's logical and what's sensible isn't always what works best.

Because on one level you already know how long your mile is: it's the distance you just walked, counting on your fingers or on your bit of knotted string. And it's the distance from your house – or wherever you started off – to the pub, or the tree with the odd-shaped trunk, or wherever it is you ended up. And for a lot of purposes it's no more use to know how many feet or how many inches that equates to than it is to know how many seconds there are in a day, or minutes in a year. You know that there's a relationship, and you sort of know that at some level it matters, but most of the time the very big and the very small seem to inhabit worlds so different that many of us can go our whole lives neither knowing nor caring how they interconnect, and be at no disadvantage at all.

We'll look at both things, but we'll do it the practical way rather than the strictly logical way – the use of it before the theory and dimensions of it – because that's how real life often works.

Anyway, you've got your mile, and let's say, for the sake of argument, that it's the distance from your house to the pub. Because you worked out the distance by counting in tens, the obvious way to divide it up would be to split it into tenths. But then you have to ask yourself how useful that would be. Once you've got the journey in your mind not as a number but as a picture of a road and houses, shops and trees, would you find it easier to think of the distance to some point on the way in terms of tenths ('I called for John six-tenths of the way to the Red Lion')? Or would you find it easier to think in terms of a simpler and more basic fraction ('He lives just over halfway there')?

Most cultures, once they've done the initial calculations, seem to find the simple half-and-half-again division most convenient, both for visualizing distances and journeys and for parcelling up tracts of land into fields and farms (even if your car mileometer gives you the distance in tenths!).

A mile is most commonly divided into halves, and often into quarters. Sometimes it's divided into eighths, but not usually any further than that. The eighth is about as long as a typical field; a sort of natural border between the world of huge things and distances and the more comfortable and familiar world of domestic-sized things, garden-sized patches of land and human-scale units.

If you are a reasonably fit person, an eighth of a mile is about as far as you can run at full pelt before your body starts giving up on you. If you are a farmer in a country where motorized tractors aren't common or practical, an eighth of a mile is about as far as your horses or oxen will pull a plough before they need a rest.

The ancient Greeks called this distance a *stadion*, and the Romans a *stadium*; and most of their fields and farms, as well as almost all of their athletic tracks, were of this length. The ploughmen of the Middle Ages knew it as a *furlong*, or furrow length, and again, shaped and sized their fields by it.

All in a day's work – the story of the acre

You trudge along the field, pushing your plough through the mud, while up ahead your oxen or horses plod and snort. And after a furlong or so you give your team a quick breather, and then you turn round and make your way back down the next furrow, and then the next, and then the next. And you do this hour after hour and day after day, and your hands get sore and your back aches. And, from time to time, you find yourself thinking, 'How much longer is all this going to take me?' And that's where the *acre* comes in, and where units like the acre come in in other countries. Because an acre, you see, is a day's work at the plough.

So how much is that? Well, if you were to have a field a furlong square, and if you were to work a five-day week, you'd have it all ploughed in two weeks. By the end of the first day, you'd have ploughed a long narrow strip, a furlong in length and a tenth of a furlong in width. And that is your acre.

The name is Old English, but comes from the Latin *ager* and the Greek *agros*, both meaning 'a field', and it's a day's work at the plough. In other countries, similar kinds of units have names that make the link with working-time more explicit: there is the almost identically sized French *journal*, and the north German *morgen* (morning) of just over half an acre.

If you want to think of it as a square, or at least a squarish oblong, the best way to visualize it is as half a soccer pitch: that's more or less an acre.

The furlong is not much used today, except by the horse-racing community. Sprinters still run the distance, expressed in metric form, but very few are aware that it ever had any relation to the mile. For most people, miles just come in halves and quarters. There's no reason why it should be otherwise: it's a free world, and measures come and go according to need and convenience. And there isn't much call for ploughing with oxen in many countries these days.

So how long *is* a mile, then?

Let's work this out. We've said a thousand double paces. A double pace is how much, exactly? You can work this out by going out to your local high street, finding a seat with a good window view in a café and watching the people walk by. What you will see is that there's something of a pattern in the distance between steps. Generally between one step and the next people leave a gap which is a bit bigger than the length of one foot, but a bit smaller than twice the length – about one and a half times the length, in fact.

So starting from the heel of the first foot they put down, there is the length of that foot, then a gap of one and a half foot-lengths, then the length of the next foot they put down, and then a gap of one and a half again before they put the first foot back down to start the next pace.

Going for a walk

Add those together and you get five. So one double pace is five feet. Which means that a thousand double paces is five thousand feet. And that's the original value of the mile.

I say 'original' because there have been one or two wobbles over the years, and over the years there has been a bit of 'inflation', for reasons that I'll come to very shortly, all of which means that the current 'official' value of the mile has ended up being slightly longer than it was when it started out.

Hup two three four . . . er, five six

A mile isn't like a foot or a yard. It exists more in your head than in your hands. You can't pick one up, or put two of them side by side to check they're both the same. Without accurate measuring equipment, what you judge to be a mile is always going to be an estimate. The count-a-thousand-paces way of estimating it is actually quite a good one, but it does demand a bit of attention and dedication. Some people are prepared to make the effort, some of the time; but not all of the people, all of the time. You might expect military surveyors employed by the Roman Empire to do it properly, but for the settled local populations, concerned more with the rough distance to the village pub or the length of time it's going to take them, roughly, to drive their herds to the nearest market town, ease and convenience have often mattered more than accuracy.

And if you're faced with two different ways of measuring out the same distance, and if one is demanding but accurate, and the other easy but not so accurate, the temptation is to go for the easy option. This, you will remember, is our great unspoken principle, the Principle of Repeated Bodges.

So if you believe that all of the fields where you live are a furlong in length, which is to say, an eighth of a mile, and if you think you have a good reason for believing this, which is that this is how far you can plough before your animals need to stop for a rest, then you can count out a mile by walking eight fields.

A choice

Except that the distance you can plough – and consequently the length of your field – varies slightly according to how good your soil is. And over time, if the technology of ploughs improves, or if the breeding stock of plough animals improves, then there will be a natural tendency for 'furlong'-sized fields to get bigger. And this means that 'miles' measured by the size of fields will tend to vary from place to place, but with an overall tendency to grow bigger, rather than smaller.

By the Middle Ages, miles across Europe were all over the place. In England, some areas still used the five-thousand-foot Roman mile but others used other 'miles' of their own devising, and it was getting out of hand. And at around this time, governments started stepping in and saying, 'enough is enough', and setting down by law what, exactly, a mile should be. But the thing was, each of these countries' official 'miles' differed from the official miles of other countries. For most people, whose travel was confined not only to their own country, but to their own little part of their own region of that country, this wasn't a problem; but for those who travelled between countries it was.

Of all the land miles created over the years, only one remains in widespread use. This is the English statute mile, and it was standardized by law not the military surveyors' way, at the 'correct' or original distance of five thousand feet, or a thousand paces, but the farmer's way, at eight furlongs. Because the accepted and commonly used length of the furlong had settled, over the years, at two hundred and twenty yards, this made the mile 5,280 feet – which is to say, fifty-six paces (or two hundred and eighty feet, or just under six per cent) longer; and this is how long it remains to this day.

If you want to know what this is in yards, just divide by three. Or rather, instead of actually dividing 5,280 by three yourself, there's a little rhyme that tells you the answer: *George the Third declared with a smile / Seventeen-sixty yards in a mile.*

George the Third: declared with a smile

It is not known whether George III ever did actually declare such a thing or, if he did, what kind of expression he had on his face as he did so, but it's a useful way of remembering a very big, awkward number. And also the date of George III's ascension to the throne. The two pieces of information are, of course, completely unrelated.

But on the other hand, if you do ever decide to use the counting-your-steps method to measure out your mile, do remember to add the extra fifty-six paces at the end. If, on the other hand, you can't be bothered with that, and if you're wearing a watch, a mile is as far as you get to in around twenty minutes, if you're walking at an average sort of speed.

Further still

If you're the sort of person who walks distances that can be measured in miles, rather than, say, the sort of person who drives to the local corner shop, or if you're one of the countless millions of human beings for whom walking is the only available or affordable way to get from one place to another, then once you get into your stride you probably wouldn't want to stop every twenty minutes. Instead, you'd probably do about an hour's worth before taking a break. Which takes us to our last measure of distance, the *league*.

The league is an ancient measure, coming originally from the Celts, and has been widely used, in various forms, over much of the world; and the simplest, most practical way to define it is as 'about an hour's walk'.

It's a useful sort of size, but knowing the exact distance you'd walk in the time does rather depend on there being a standard measure called an hour, and on your being able to measure it accurately, which before the invention of clocks and watches was presumably a little tricky. It also varies according to how fit and how fast you are, and what sort of terrain you're walking over, and so it has tended to be a rule-of-thumb unit rather than a scientifically precise one. Where it has been fixed and codified, the league has traditionally been set at three miles, using whichever local version of the mile is in use. This means that the Portuguese version, the *legoa*, at three *milhas*, works out at almost four statute miles, or an hour's very brisk walking; whilst the Spanish *legua* of five hundred *varas* is a lot shorter at just over two and a half miles, or an hour's leisurely post-siesta stroll.

A Collection of Kitchenware

After all that walking you could probably do with a drink. So why not pour yourself one?

But as with everything in this book, there is no such thing as a free lunch. Or even a drink that you don't have to work for.

Because your task now is to come up with a complete system of fluid measure, while you're getting yourself that drink.

Now, I can't see what you've got yourself, being as how I'm sitting here writing in a different place, and at a different time, to where you are now. But my guess is that you've got yourself either a *regular*-sized drink, in some-thing the size of a teacup or a coffee cup, or a *medium*-sized drink in a mug or a soft-drink can, or else you're really very thirsty indeed and you're going to gulp down the entire contents of a *large*-sized glass or even a bottle.

Almost all of the cups, glasses and cans we use to drink from fall within the narrow range of volumes covered by this 'regular, medium and large' range. This is as true of cups and glasses in metric countries as it is in non-metric ones: basic appetites are basic appetites, whichever system they are measured in.

Some countries do use smaller cups – espresso cups, for example, or shot-glasses – but they are more about getting a shot of caffeine or alcohol than they are about quenching your thirst. However, there aren't very many extra-extra-large sizes in common use, designed to be drunk by one person in a single sitting, not unless you count those great buckets of cola drunk in cinemas in certain countries by people who are clinically obese, or well on the way to becoming so.

Let's look at some proportions now. If you get the biggest container you'd want to drink all of the contents of, if you were really thirsty, and poured the contents into regular-sized teacups, how many would you fill up?

The answer will vary slightly from person to person, depending on your age, your sex, your size, and how hot or cold a climate you live in, but the average number of 'normal' cups that a big drink fills up is two.

Going down in size, you will find that half a 'normal' cup is – well, half a cup, or about as much as you'd pour into an average-sized glass of wine. Divide that in half again and you'll reach the size of an espresso cup (if you own any), which is about as small a drink as most people would still class as a drink, and not a 'gulp' or 'taster'.

Divide your espresso cup in half and half again and you'll fill up a tablespoon.

What we seem to be arriving at here is a sort of natural binary system, with the sizes appropriate for most everyday needs and appetites being most efficiently 'marked off' by a range of containers each roughly half the size of the one above, and twice the size of the one below. The only real exception is the soft-drink-can size, which has evolved over the years to a capacity of one and a half standard cups, which means that it fits exactly halfway between a regular drink and a large one. This makes it just the best size to sell to anyone, anywhere, who has any kind of thirst: neither too big for someone with a casual thirst, nor too small for someone with a big thirst. That's probably not an accident.

What we're *not* arriving at in our collection of kitchenware is a natural decimal system, with each container we own being ten, or a hundred, or a thousand times bigger or smaller than the next.

This rough 'natural binary' also works for quantities bigger than the amount you'd drink in one go. If you pour out the contents of your large drink, you will find that roughly two of them will get you to the size of the kind of bottle or carton that you can carry and handle with ease, and store comfortably in your cupboard or fridge.

Two bottles or cartons will fill up a regular-sized mixing bowl or saucepan, and two of those will fill up a large mixing bowl or saucepan, or a small bucket.

The diagram below shows a typical range of the kinds of containers that people use to drink from, or to cook with in their kitchens, arranged in order of size.

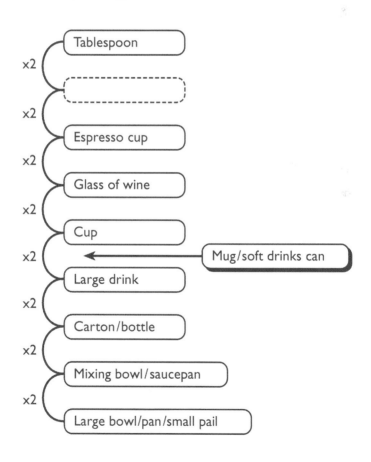

If this is the order of size, the question now is, what are the *actual* sizes, and why? And that's what we look at in the next section.

The Metric Pint

Let's look at some sizes and proportions now, and let's do a little experiment. You may enjoy this, or else you may find it somewhat disgusting, depending on your disposition and oral hygiene. What we want to do is to discover the relationship, if there is one, between the sizes of the handy-sized containers in your kitchen, and your own body's natural capacities.

Your body has a number of natural 'receptacles' to hold liquids, but we'll give the stomach a miss here, both because it's fairly elastic and able to hold either a little or a lot, and because it's not exactly easy to use it for measuring – you won't want to be filling it to capacity and then emptying it in a hurry, just for the sake of working out how big your pots and pans are. Likewise, you *could* fill your bladder and use that, but it's not exactly quick or practical, most of the time. Or, indeed, pleasant.

So we'll have a look at how much liquid your mouth can hold, and see what we can learn from that.

To do what we're going to do, you will need your cups and glasses, a large jug and some water. Begin by pouring yourself some water, and then sip from it, without swallowing, until you just reach the edge of feeling pleasant or comfortable. Don't force it – not yet, anyway. Then spit it out into a cup or jug and see how much you've got. For an average adult, a 'comfortable' mouthful is about an espresso cup's worth.

Now take another drink, but this time really try to take as much water as you absolutely can, until you can't fit any more in and it begins to hurt the back of your tongue. Then spit that out. What you should find is that you will have roughly twice as much as you had the first time, and it should either fill two espresso cups or else half fill an average-sized teacup. One more mouthful like that and you should fill your teacup more or less to the brim.

Then take the water from that cup (you might prefer to replace it with clean water first) and pour it into one of your bigger receptacles. What you should find is that there is a simple relationship between the amount you can hold in your mouth and the sizes of the array of cups, glasses and jugs that you find useful in your kitchen.

Your large drink in particular is an interesting thing, because not only is it roughly four times as much as you can fit in your mouth, but it's also, more or less, about as much as you can fit in your bladder. Or, to put it another way, the human bladder holds about four complete mouthfuls. I suggest you just take my word for that.

The other thing about the four-mouthful or two-teacup measure is that if you weigh it, you'll find that you will have roughly a pound, or four hundred and fifty grammes, of liquid.

For a long time, throughout much of the world, a pound of liquid was a basic unit of fluid measure. In England, the exact same word was used for both, although over time the way the word 'pound' was pronounced for liquids evolved into a new word: *pint*. But a pound was a pint and a pint was a pound. The Americans, who still use the English system, have a little rhyme for it: *a pint's a pound the whole world round*.

The French, meanwhile, had a unit of almost exactly the same size – just ever-so-slightly bigger – called a *chopine*; but although they are an artistic nation, and rightly proud of the sophistication of their high culture, it is not known whether they also have, or had, such a poem about *their* national drink-sizing.

Anyway, this pint-sized pound (or pound-sized pint) was divided up in exactly the same binary way as a regular pound, which is to say, into halves over and over again;

and many of the divisions, through long usage, came to have their own names.

Half a pint is a *cup* (a *demiard* in France).

Half a cup is a maximum-mouthful-sized *gill* (pronounced 'jill', from the Latin *gillo*, meaning a small water-pot or wine vessel), an amount known as a *posson* to the French and as a *wineglass* to American bartenders. It would be plain wrong to say that there's any connection whatsoever between a *gill* and the expression 'full to the gills' (as in fishes' gills, with a hard 'g'), but that's pretty much how you feel if you tried to hold one in your mouth. Half of that is a comfortable mouthful-sized quarter cup (a *roquille*); and half of that is an ounce. And half an ounce is a tablespoon.

Going upwards from the pint, two pints make a quarter of a gallon: a *quart* to the English (or, slightly confusingly, a *pinte* to the French).

Double that and you get half a gallon. And double that and you get a small bucketful called a *gallon*, from the Latin word *galeta*, meaning 'a small bucketful'.

That was the way things were.

And then two things happened – at around the same time – that tempted two nations to mess with it. The first thing was that the French had a revolution. The second thing was that the British lost an empire in the west (under

King George III, who, you will remember, was known for declaring with a smile) and gained a new empire in the east.

In France, the revolutionaries wanted to sweep away everything to do with the old regime. They didn't just want to change the way the country was run: they wanted to transform beyond all recognition the way people lived and thought. People were going to turn away from tradition and superstition and live instead in the bright light of

The French, having a revolution

Reason. To help bring this about, everything was to be created anew, right down to the calendar, the clock and the way people weighed and measured. At the heart of this new way of weighing and measuring was a new 'scientific' unit of length, called the metre; and – as volume is length, cubed – the rational choice for a new unit to replace the old chopines and pintes used before the revolution was the metre, cubed. This new unit was called the *stere*.

However, if you've ever spent much time looking through dictionaries, you'll have noticed that 'rational' and 'good' tend to be listed separately, as different words. There is a reason for this.

There was a problem with the stere. The problem was that no one actually liked it or wanted it. It's a massive measure, far too big to drink and far too heavy, even, for most people to carry. So although the new metric system was strictly enforced for a time, and although working in the old system was seen as a 'thought-crime' that marked you out as a counter-revolutionary, the truth is that because of the stere it met with far more hostility (or shoulder-shrugging indifference) than had been expected. If anything, this hostility increased over time, and as the initial fervour of the revolution lessened, so people quietly got back to using the weights and measures they were most comfortable with.

By the early nineteenth century the metric system was unloved and unused, and Emperor Napoleon decided he had had enough of it. He had never been the greatest enthusiast, anyway: he refused to learn how to use it himself, and it is recorded how on a visit to a gunpowder factory in the town of Essonnes he quizzed the plant's manager in great detail about the manufacturing process, and then, every time he was given an answer in kilogrammes, he insisted on having it converted to the traditional *livres poids de marc*.

'Nothing,' he said, 'is more contrary to the organization of the mind, memory and imagination. The new system will be a stumbling block and source of difficulties for generations to come. It is just tormenting the people with trivia.'

So it wasn't a huge surprise to anyone when in 1816 a new law was passed officially allowing the return of *mesures usuelles*, which is to say, the old pre-metric measures, but slightly amended to their nearest metric value and nominally reclassified as 'metric'. The old fathom, or *toise*, came back as a unit of two metres; the livre came back as a 'pound' of five hundred grammes; and the old pinte, or French quart, came back as a measure called the *litre*. This litre was officially defined as one cubic *decimetre*, but in size it was so close to the pinte that for most purposes people could, and did, use their old pinte measuring jugs. More than this, the litre was divided in exactly the same way as the old pinte, into halves (known as *demis*), quarters (*quarts*), eighths (*huitièmes*) and sixteenths (*seizièmes*).

When Napoleon went, his reforms went, the government that replaced him disapproved of the French using the 'backward' and 'superstitious' measures of the past (even dressed up in 'metric' clothes), and in 1825 they enacted new legislation outlawing them again, and compelling people to use the 'pure' metric standards. Except one thing was different: no one, not even the government, wanted the stere back. Logical it may have been, and properly, seriously metric, and just the thing for a system based around the metre; but it was just far too big to be of any use. So the stere stayed dropped, and the metric quart, or litre, stayed instead, and remains in place to this day. Although not *officially* so – as we'll see a little later.

So that was France, and that was the metric system. But even as the metric system took its first big step in evolving towards the traditional values of the old system it had replaced, the system used by the British took a step in a slightly different direction.

As the British Empire grew, so did international trade; and as international trade grew, so did the need to load barrels and bales onto ships, and with that the need to know how much weight you were putting into your hold, so as not to overload your boat and cause it to sink.

The British, getting an empire

Because a pint weighed a pound, and a gallon of eight pints weighed eight pounds, the weight of cargoes of barrels generally had to be worked out in multiples of eight. I don't know how well you know your eight times table, but my guess is that your knowledge would be sorely tested if you had to calculate whole shipfuls of eights. To sort this out there were a number of things that could have been done. The most obvious course of action would have been to make export barrels in multiples of ten pints. That probably would have done it.

But at this time – the early 1820s – there was a lot of interest in what was going on across the Channel, with the French authorities getting ready to reinstate their metric system; and it struck someone, somewhere, as rather a good idea to change the size of the gallon, so that it represented not eight pounds of liquid but ten. Of course, the barrels would still have to be made bigger as well, so it wouldn't actually save any work at all, but there would be the added bonus, as they saw it, of the gallon becoming more modern and cosmopolitan in the process.

So in 1824 a new law was passed setting the size of the gallon at the volume of ten pounds of pure distilled water, rather than the old eight of wine. But because the number of pints to the gallon wasn't changed by the new law, this meant that the pint itself had to get bigger, growing from sixteen ounces to twenty, which is a pound and a quarter, or five full mouthfuls instead of the old four. This is actually rather a lot to drink in one go, for a lot of people – although beer-drinkers might argue otherwise. The half-pint or cup, meanwhile, grew from eight ounces to ten; which, again, is actually a bit bigger than a real cup that you might drink your tea from. And perhaps because of this, the use of cups to describe the size of a . . . well, a cup, began to die out on this side of the Atlantic.

The British get a bigger pint

The Americans, meanwhile, had long been independent from British rule, and were not subject to any laws on weights and measures dreamed up by some bright spark in London. They quietly declined to join in with this 'modernizing' exercise, and kept their pints as pounds and their cups as cups, and maintained the binary divisibility of their pint right down through cup, half-cup, quarter-cup and ounce, right down to the half-ounce tablespoon size.

'Two nations separated by a common language'

British and American weights and measures come from the same roots and remain pretty much identical in most respects today. They do have different names, though.

Because the American units came over with the English settlers, the Americans call their version the *English system*, even though they themselves are no longer English. Because the British standardized their units in the 1820s to cope with the demands of trade with their growing empire, they call their version the *imperial system*, even though they no longer have an empire.

There are one or two areas in which the two systems differ, though. In each case the reasons for the differences are the same, and are to do with each nation's size and situation. Britain is a small island nation, and has always depended on international trade. The USA is a large self-sufficient nation, able to please itself.

Fluid measure we know about already: the Americans have a sixteen-ounce pint and the British a twenty-ounce one. More than this, the volume of a fluid ounce is actually ever so slightly different in each of the two nations. They're both the same in that they're the volume occupied by an ounce of fluid, but they're based on different fluids. Different fluids occupy different volumes depending on something known as *specific gravity*. Imagine an ounce of molten lead and an ounce of whipped cream and you'll get the picture. The American fluid ounce is an old trader's measure based on an ounce of wine, as was the British version until they decided to go all 'scientific' in the 1820s, and said that their fluid ounce would henceforth be based on distilled water at a specified temperature and pressure. The difference between the two ounces isn't a very big or noticeable one, but it's a difference all the same.

There is another unit that differs: it is a unit of weight called the *hundredweight*. It's a multiple of the pound. So here's your question: how many pounds do you think ought to be in a hundredweight? No, that's where you're wrong: it's a hundred *and twelve* pounds, actually – or at least, that's what it is in Britain, and has been so since the fourteenth century. The reason for this is that the continental Europeans used an ancient measure of about that size called the *quintal* or *zentner*, and as trade across the Channel increased, the hundredweight was adjusted to match it. For the Americans, meanwhile, it just seemed a bit daft to have a hundredweight that wasn't a hundred of anything, and so they changed it back again. This means that the *ton*, which is twenty hundredweight, is two thousand pounds exactly in America, but more in Britain. There's also a metric ton, or tonne, which is roughly halfway between a British and an American one. As far as the average person in the street is concerned, though, they're all much of a muchness, and all much heavier than anything you'd ever want to try to pick up and if any one of those real tons were to drop on that person –

say in the form of a concert-sized grand piano left suspended by careless removal men using a frayed rope looped over a pulley-wheel hanging from a third-floor window (or indeed a fourth-floor window by the American numbering system), then the results would be much the same.

This is how things remain today. But whether they continue to remain that way remains to be seen, because as I write, two big changes are under way.

One of these changes involves the metric system as it is used in practice and the other involves the imperial system. It is hard to tell quite where they will end up, or how long it will take, but here is what is happening.

The metric system, as used in practice, now seems to be re-embracing the use of cups. In 'metric' Australia, it is now more common for the recipes published in books and magazines to show fractions of a litre in cups than it is to see them in millilitres; and this is a trend that seems to be spreading. There are some variations, but a metric cup is typically sized at two hundred and fifty millilitres and a half-cup at one hundred and twenty-five – the same sizes, in fact, as the old *mesures usuelles* divisions of the litre. This adds an extra dimension to the French 'metric quart' of the 1820s, and puts its volume at eight metric cups, or sixteen metric half-cups. Compared to, say, a non-metric quart of eight non-metric cups, or sixteen non-metric half-cups. And it is a very short step from there to a metric two-cup measure (which you might call a metric pint, or demi) of half a litre. But of course it would be a slightly different size from the non-metric version, so that might make it more acceptable to the sort of people who like metric because they think it's 'more modern' and 'more international' and 'not provincial'. The chart opposite shows just how different it would be.

Which, compared with the original English quart, pint and cup measures, isn't very different at all.

How it's all going to turn out in the end only time will tell. Perhaps the litre will go further in its transformation into a full-on quart; perhaps cups and fractions of cups will displace millilitres for more and more everyday purposes, and will acquire a pint measure as well. On the other hand, perhaps it won't.

And after two hundred years, the British seem to have grown quite fond of their twenty-ounce pint, and find it particularly useful for beer and milk. This may be just familiarity and nostalgia, or it may be that its position nearly midway between the old English pint and the quart fulfils a function, allowing it to be drunk entire by a person with a very big appetite, or else split into two or more reasonably large drinks. Like the one-and-a-half-cup

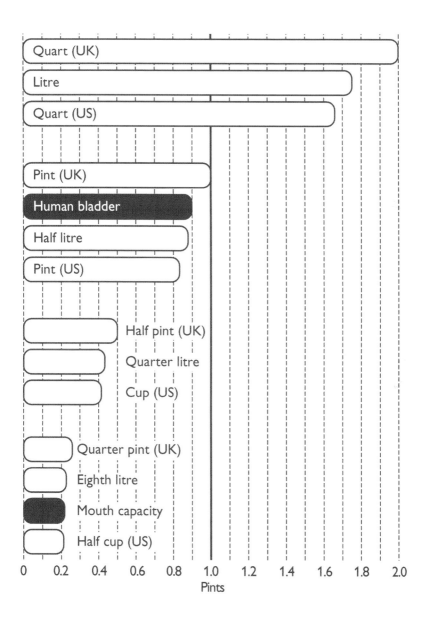

soft-drink-can size that sits midway between the cup and the pint, the twenty-ounce pint may yet turn out to have been a good idea, after all. But then again, it may not.

The wheel, as it were, is still in spin. And there's no telling, at this stage, quite where it will end up.

The Future

We've spent most of this book looking at how the simple bodges and rules of thumb used by ordinary people have ended up creating whole systems of measurement that have endured for hundreds or even thousands of years.

But a lot of people nowadays would argue that, while all this sort of stuff may (or may not) be interesting, none of it really matters any more, because the world has changed and people have changed. The world has changed, they'd say, and people have changed, because of the metric system, which has brought about what you might (or might not) call a *paradigm shift* – which is to say, it's brought with it a whole new way of looking at the world, which makes all previous ways obsolete.

What it is, they'd say, is *progress*.

In metric, the argument goes, we've got a simpler, more rational and generally *better* way of measuring the world, a way that people in the olden days weren't clever enough or advanced enough to grasp.

And because of this, people think that the spread of metric is *historically inevitable*, which is to say that it will soon be everywhere, and used by everyone, and that this will happen within the space of a single generation, as the remaining native users of the 'old' systems die out.

So it seems a good time now to take a look at metric, to see whether we really have reached the end of history, as far as measurement is concerned, and to see how fit it is, and how likely it is, to be the be-all and end-all of all sensible measurement from this time forward.

The first thing to say is that there isn't just one, 'metric system'. There are actually quite a few of them, or, if you prefer, quite a few variants of the system that came out of the French Revolution. But there is only one of them that holds itself to be the official, genuine version.

This official, 'genuine' metric system is called the *Système Internationale*, or SI. SI is hardcore metric, or scientists' metric. There's an international committee of experts that runs it, set up in the 1960s, and their main concern is to keep it pure, accurate, and up to date with the latest scientific developments.

What matters most of all for the international committee of the *Système Internationale* is that their system remains spare, consistent, modern and logical. This is not always the same as what matters for carpenters and stallholders and people like that; but that's not really the concern of the SI committee, and nor should it be: they're there to keep things working properly for the scientists.

For a carpenter, it may well be useful to have a metric 'hand' – the ten-centimetre decimetre; and for a stall-holder it might be handy to have a measure of the weight of a handful of sausages – the five-hundred-gramme livre or gongjin; but for the purposes of science, these sorts of extra measures overcomplicate things, and, all in all, are just extra sources of potential error.

For this reason, one of the first things the SI committee did after it was set up was to abolish or downgrade a whole list of measures it felt to be impure or unnecessary. There was thought to be no need for the decimetre, or the centimetre, and so they both went. Likewise the hectare, a measure of area used by farmers to divide up land. The litre didn't fare any better: it came off the list of approved units, to be replaced with the cubic metre. The cubic metre, you may remember, used to be called the stere, back in the early days of metric, when it was rejected by everyone including Napoleon.

Litre 'too big' shock!

Metric purists had long felt that the cubic metre was a purer and more rational unit of volume than the litre; and they were doubly unhappy when it was discovered that the 'standard' litre in Paris was actually 0.000028 cubic decimetres bigger than it ought to have been. Which is to say, each side of the precious metal block was 'wrong' by slightly less than a thousandth of a millimetre.

There are several things they could have done about this.

They could have done the pragmatic thing and just ignored it – the difference was extremely small and, right up until that point, no one had ever noticed it.

They could, on the other hand, have done the practical thing, and got a man in with a sheet of emery paper, or whatever it is that standard-litre repairmen use, to sort it out.

Or else they could have taken logic and rationality way above and beyond the call of duty, and concluded that since the litre in their possession was the official standard one, and since that official standard one was the size it was, rather than the size it ought to have been, then they were duty bound to redefine the litre, so that all other litres in the world should be adjusted to match the error in their one. And this is what they did, and this is why, for much of the twentieth century, the official definition of a litre was '1.000028 cubic decimetres'.

> After sixty years or so the embarrassment of it got too
> much for them and they set it back to exactly one again;
> and then, a few years later, with the advent of the Système
> Internationale, they did away with the litre altogether, and
> the decimetre too, for good measure.

There was another way in which the SI committee looked back to the early days of the metric system: they turned their attention to time, and how to measure it.

The early revolutionaries had tried a short-lived experiment in decimal time, with a ten-hour day and a ten-day week. That failed pretty quickly, not least because it took a lot longer for the weekend to come round, and there were fewer of them to the year, and people felt cheated.

Two hundred years later, the SI people had another go at rationalizing the clock and calendar, for scientific purposes. What they did this time was to abolish or downgrade *all* the non-metric measures of time – which, actually, is all of them. They did keep the *second*, though, but only after they redefined it as '9,192,631,770 cycles of the vibration of the caesium-133 atom'. This probably doesn't mean a great deal to the average person in the street, but it is – or was – probably the most precise and reliable way of defining the basic unit of time that science could devise. And should an even more precise and constant way come along at some point in the future, then the chances are that the SI committee will get round to redefining the second to the new standard.

All of which are perfectly fine, and perfectly right and proper things for a scientific committee to do. That's what they get paid for. But it's just that, scientist or not, you wouldn't necessarily want to apply strict SI to the rest of your life outside the lab. And you could probably count on the fingers of no hands the number of university research physicists who measure the time until their drink with their colleagues in the campus bar in *kiloseconds*, or who order their beer in precise decimal fractions of a cubic litre.

*

At some point, we all step back into the everyday world, where everyday purposes call for everyday measures, and where the everyday measures we find most useful are the ones that relate to the dimensions, the appetites and the purposes of ordinary people like you and me.

It's perhaps pushing it a bit far to say that there's a 'natural' system of weights and measures built into humanity, but if there's not a fully formed system, then at the very least there's a clear and well-defined set of natural boundaries; and all 'everyday' systems need to be based within these boundaries, if they are to survive and prosper.

Anywhere the metric system has been in common use for any appreciable length of time, 'everyday' metric units have emerged, and the longer the common people have had to get their grubby hands on them, the more the Principle of Repeated Bodges has creept in, and the less ideologically pure and SI-like they have become.

In some places the metric system is still reasonably close to the 'official' version, save for the use of centimetres, hectares and a few other 'old' metric units, but in others new local measures have been added, or old *mesures usuelles* retained or redefined, like five-hundred-gramme livres and gongjins, and like 'metric cattys' and seventy-centimetre braccios. Some of these units are nominally metric, but others – like the braccio – are hardly even that. Some, like the inches still used by Dutch and German plumbers, never even went metric in the first place – because ideological purity is no substitute for being able to stick your thumb down a pipe to stop the water flow.

Some 'impure' measures have no formal name or acknowledgement. In some countries, people are unaware even of the fact that they routinely divide the kilogramme into unnamed multiples of twenty-five-gramme 'chunks' in their minds and in their shops; in other countries, new units such as the fifty-gramme liang have become formal, government-accepted units.

Wherever you look, the versions of the metric system used in everyday life have been customized and bastardized to suit the needs of people who don't always have a string of letters after their name, or a scientific calculator in their top pocket. Metric cooks have started to work in cups; metric timber and other building materials across the world are increasingly sold in multiples of thirty centimetres, which – as it happens – is almost exactly the length of a builder's twelve-inch measuring boot.

So there does seem to be some sort of 'historical inevitability' at work, but not in the way that a lot of people think. What seems to be inevitable about the kinds of weights and measures used by normal people is that, in the long run, they all end up either adapting to give people what they want, or else dying out altogether. The more that different systems change and adapt to meet the needs of their users, the more they become the same. And the more they become the same, the more they become like the traditional systems which, almost by definition, *must* have got something right, or else they wouldn't have survived for as long as they have.

This is not to say that traditional systems are *always* right, that they are right in every aspect, or that they don't go off down obscure blind alleys from time to time – it's just that over time, and with lots of use by lots of people whose main concern is to just have the easiest and most practical toolkit for everyday life, and not to have to think too hard about it or work too hard at it, these things will tend to get ironed out; and any measures that stay tied to old activities that no one does any more end up getting quietly dropped and replaced with newer ones that relate to newer purposes. An urban world geared around technology and international trade may have less need of the barleycorn and more need of new measures like the *TEU*, a 'twenty-foot equivalent unit', which is the twenty-by-eight-by-eight-and-a-bit-foot volume of a cargo container, in which freight ships (and increasingly trains and trucks) are measured.

There is a place in this world for precise scientific measurement based upon the vibrations of atoms and the wavelength of light, and there is a place in the world for the everyday measures of normal people, measures based upon workmen's boots, and carpenters' thumb-widths, and all the twists and turns and knots and oddities of the crooked grain that is human nature. And anyone who tries to wish away or legislate away the one in favour of the other is, in the end, doomed to fail.

In measurement, as in so much else, the only right and sensible thing for any of us to do in this life is to live and let live.

Glossary

SYSTEMS OF MEASUREMENT

Traditional systems

English system – Traditional system of measures used in America, but not in England. Uses cups (eight fluid ounces) for cooking. Has a pint that weighs a pound and a hundredweight that weighs a hundred pounds.

imperial system – The British system of measures. Almost identical to the English system, but with a couple of differences designed for overseas trade, both with continental Europe and with an empire. The main ones are a gallon that weighs ten pounds (to make it easier to calculate ships' cargoes), which results in a twenty-ounce pint. Also has a hundredweight of a hundred and twelve pounds (to make it comparable with the European quintal) and a fourteen-pound unit called the stone that British people measure their weight by (because they do).

Eastern systems – A group of systems based on traditional Chinese measures. One of the most intact and widely used versions is the Japanese Shakkanho system, which has been used continuously since the seventh century. In theory, it was officially replaced by the metric system in 1924, but it had to be officially replaced by the metric system again in 1966, because no one seemed to have noticed first time round. Despite this, large areas of Japanese life continue to operate in Shakkanho. And although, in theory, it's against the law to use it for official purposes, it's been creeping back even there, with the 2005 census allowing people to give the size of their properties in traditional units once more.

For length, the system uses a kanejaku or 'carpenter's measure' just short of an English foot (11.93 inches), for volume a sho just short of half a US gallon (sixty-one US fluid ounces), and for weight a kin of one and a third pounds, or six hundred grammes, which comes from the Chinese jin, and which is a very common size for a unit of weight in Asia. The Japanese also have a traditional sake-cup size, the go, which is just over six ounces (or half a soft-drinks can).

Metric systems

Système Internationale (SI) – Fiercely rational system of measures used by scientists. Managed by an international committee of the great and the good, who devote a lot time and energy to abolishing measures they disapprove of, or deem to be impure or unnecessary. So far they've issued prohibitions against the centimetre, the decimetre, the hectare, the centilitre and the litre, to name but a few. Though committee members have been spotted wearing regular wristwatches with minute- and hour-hands, the only measure of time that has full SI recognition is the second.

metric – A generic term covering a range of more easygoing (or less 'pure') versions of the system of weights and measures that came out of the French Revolution. Everyday metric measures today typically include most of the SI measures, together with some 'old' metric units such as litres and hectares and occasionally a number of dodgy, nominally metric local concoctions with names such as metric liang and braccio.

Changes between systems

serial metrication – Serial metrication is what happens when a government decides to change the way a nation weighs and measures, but no one takes a blind bit of notice, and so the government feels compelled to repeat the whole process all over again a few years later, and perhaps to threaten more dire consequences for non-compliance. France was the earliest serial metricator whilst Japan, which has been trying since 1924, is probably the longest running. Its neighbour South Korea had an almost completely unsuccessful attempt in 1961, which was followed up in 2007 by a second go, whilst over on the other side of the world Guyana has probably had the greatest number of attempts over the shortest space of time, with official switches every few years since 1981.

LENGTH

Tiny lengths

millimetre – About as thick as a credit-card. The littlest 'normal' unit on a metric ruler.

barleycorn – Old English measure of a third of an inch, the length of a seed of barley. Still used for shoe-sizing in Britain and America.

centimetre – Smallish unit of ten millimetres. Not an official subdivision of the metre any more. But people still use it, anyway.

Hands, fingers and thumbs

inch – Make a mark, or knock in a nail, on either side of the knuckle of your thumb, and that's your inch. In most languages apart from English, the word for inch is the same as the word for thumb. Twelve of them make a foot.

hand – Four inches – the width of a man's hand at the widest part, where the thumb joins the palm. Also the width of a brick, which people have been known to pick up with their hands.

decimetre – A hand by another name. Ten centimetres – which is four inches, as near as dammit.

Feet, arms and legs

foot – The length of the sole of an average man's shoe.

metric foot – Thirty centimetres, the metric equivalent of the traditional foot. This is The Foot That Dare Not Speak Its Name. It exists, in the form of thirty-centimetre rulers, and the sale of timber and building materials in thirty-centimetre units, but it just doesn't sound very metric to call it a foot, and so, for the moment, it doesn't have an official name.

braccio – An Italian measure the length of your arm. Now standardized as an informal 'metric' measure of seventy centimetres, or twenty-seven and a half inches.

yard – Three feet. The size of a stick the length of a man's leg, from ground to hip joint.

metre – The measure the metric system is named after. It's had a number of 'official' definitions over the years, ranging from one ten-millionth of a quadrant of earth's circumference, as measured along an imaginary line passing through Paris to the North Pole (except that they got the distance wrong), to something else to do with wavelengths of light. To you and me, though, it's pretty much the same thing as a yard, but about three inches longer. This means that a metre stick, used as a walking stick, would reach to somewhere around your belt, rather than just to your hip joint.

fathom – The width of a man's outstretched arms. Also the depth of water in which it's too deep for most of us to stand on the bottom without drowning.

Hence it's mainly used by people who go out in boats. Known by a variety of names around the world, including favn and famn in Scandinavia and ken in Japan.

The French modernist architect Le Corbusier called it 'an English height', on account of reading about 'six-foot tall' characters in Sherlock Holmes novels, and he based his Modulor system of architectural proportions around it.

WEIGHT

Tiny weights

gramme – A very small weight – one thousandth of a kilogramme. From the Latin *gramma*, meaning 'a small weight'.

grain – The weight of a grain of barley, or barleycorn, which the Anglo-Saxons had rather a lot of, and did rather a lot with. When they weren't eating them, they used them to weigh and measure with. Probably they used them for entertainment, too. There are seven thousand grains in a pound, but you probably wouldn't want to sit there counting them all out to weigh up your turnips, or whatever it was that the Anglo-Saxons ate with their barley.

Pebbles

ounce – A pebble-sized weight. A sixteenth of a pound.

tael, tahil or liang – In the West, we have pounds, which we divide by sixteen to make ounces. In the East, they have cattys and jins, which are a bit bigger. They divide these into sixteen, too; and what you get from that is a tael or tahil, or a traditional

Chinese liang. So it's a pebble-sized unit, a bit like an ounce but about a third bigger.

metric liang – Chinese metric unit of fifty grammes, one tenth of a five-hundred-gramme gongjin.

Handfuls

pound – The weight of a comfortable handful of apples. The 'lb' abbreviation comes from the Latin *libra pondo*, meaning 'pound of weight'.

livre/jin/Pfund/gongjin – Traditional names formerly used for local 'pound' measures, now used to describe half a kilogramme (equal to about eighteen ounces).

catty/kin/jin – Units of weight used in eastern Asia. Typically one and a third pounds, or six hundred grammes. A maximum-capacity handful of apples.

Small bagfuls

kilogramme – The weight of a litre of water. The amount of apples which, if you try to hold them all, you start dropping them all over the floor, even if you use both hands. Or at least, that's what it is in my experience.

Bigger bagfuls

stone – British unit of fourteen pounds. Mostly used for weighing people.

Heavy weights

hundredweight – A hundred pounds, if you're American; a hundred and twelve, if you're British.

ton – Twenty hundredweight: 2,000 pounds in America, 2,240 pounds in Britain, heavy enough to squash you in either.

tonne – A metric ton of a thousand kilogrammes (2,205 lb). Heavy enough to squash a Frenchman.

VOLUME

Tiny volumes

millilitre – Also known as the cubic centimetre, the millilitre is a thousandth of a litre.

Spoons and cups

teaspoon – A third of a tablespoon, a sixth of a fluid ounce, or five millilitres; you can also stir your tea with it.

tablespoon – Half a fluid ounce, three teaspoons, or fifteen millilitres; you cannot stir a table with it.

fluid ounce – The space taken up by an ounce of water, which is two tablespoons, or about half an espresso cup. Because liquids have different densities and expand and contract slightly according to temperature and pressure, and because the British and the Americans weren't really speaking to each other at the time the fluid ounce was formalized, a slightly different standard is specified on either side of the Atlantic. This means that the American fluid ounce is ever so slightly the larger, but not by so much as you'd really notice.

gill – A quarter of a pint, or half a cup. From the Latin word *gillo*, meaning a small water pot or wine vessel. Also about as much as your mouth can take before you feel absolutely, painfully full to the gills.

cup – An American measure of half a US pint, or eight fluid ounces. The Canadians use the cup, too, despite the fact that they like to maintain a degree of distinction from their neighbours and, by and large, they try not to use American measures.

metric cup – A measure of a quarter of a litre, or two hundred and fifty millilitres, being the nearest metric equivalent to the US cup. Widely used in Australia.

Bottles and buckets

pint – A large drink. The name comes from the same root as the word 'pound'. In the USA a pint is sixteen fluid ounces, and weighs a pound. In Britain, it's twenty and weighs a pound and a quarter.

quart – Two pints, or a quarter of a gallon.

litre – An old French measure, more or less the same size as a US quart, which was abolished, then reborn and renamed, and then abolished again – although no one seems to have taken much heed of that last part.

Originally called the pinte and equal to 952 millilitres, it was brought back, rounded up a touch to 'one cubic decimetre' and renamed the litre (taking its name from the old French dry volume measure the litron) under Napoleon, as a replacement for the unloved cubic metre or stere.

Metric purists were never altogether happy with the litre, particularly when they found out that the 'standard' litre in Paris was slightly too big; and it was dropped from the official SI metric pantheon altogether in the twentieth century. Although the cubic metre was reinstated as the official unit of volume, soft-drinks manufacturers have yet to start selling their wares in cubic-metre bottles.

sho – Traditional Japanese measure more or less the same size as half a US gallon.

gallon – Four quarts, or eight pints. A small bucketful. From the Latin *galeta*, meaning 'a small bucketful'.

DISTANCE

Short distances

furlong – An eighth of a mile, which is two hundred and twenty yards, or just over two hundred metres. Roughly the distance you can run full-pelt before your body begins to pack up on you, and also the distance a team of horses or oxen can pull a plough before they need a bit of a breather. The ancient Greeks and Romans used similar measures called the stadion and stadium, respectively, which were typically a little more than two hundred yards, and a little less than two hundred metres.

chains, rods, poles, perches etc. – For the purposes of parcelling out land, the furlong has been divided up into a number of different subdivisions over the years. In the 1620s, long before the event of those tape measures used by grounds-men on cricket-pitches, the mathematician Edmund Gunter developed a tool for surveyors which consisted of a standard-sized roll of chain. The idea behind chain, rather than rope or string, is that chains don't stretch in wet weather, and so they remain accurate.

Now, a whole furlong's length of metal chain would be a complete pain to lug around with you, and so they settled on a smaller length of four rods. We'll come to rods in just a moment, but for now it's enough to know that four of them, the length of one of Gunter's chains, equals a tenth of a furlong, or twenty-two yards, which seems as good a division as any; and this distance became known as a chain.

Now for the rod. Or pole. Or perch.

There's a little story about how these came to be so called. In Saxon times, and through the Middle Ages, fields were often ploughed by a team of four oxen controlled by a small boy with a big stick, and this boy's stick would have to be big enough to reach all four of his animals from where he was. That stick was known as a rod (or goad, or gyrd, meaning 'straight stick', from which we also get the word 'yard') if he were a Saxon. Or else it would be called a pole, or a perch if he were Norman.

Sometimes these sticks were used to measure distances with.

Or so they say.

But a rod is five and a half yards long: over sixteen feet. That's not a 'rod', it's a tree! And for a boy – or even a man – to casually wander along carrying something that size in one hand while he steadies the plough with the other, well, he must have been eating rather a lot of barley and turnips, is all I'd say. Or is there something I'm not understanding?

kilometre – A thousand metres. Equal to around five furlongs (five-eighths of a mile).

mile – Originally the *mille passus*, a Roman measure of a thousand paces (or two thousand individual steps; a pace being the distance between successive falls of the same foot). An average of five feet per pace makes the original *mille passus* five thousand feet. Centuries of use and abuse brought in all sorts of variations between one country and the next (and, more than that, between one region and the next and even one town and the next), with the result that governments felt compelled to act. In the 1590s the English government looked at all of the miles on offer and decided to standardize at eight furlongs, since furlongs were an important agricultural measure at the time. This means that the English statute mile gained an extra fifty-six paces, and ended up at 5,280 feet or 1,760 yards.

league – Three miles: about as far as you'd walk in an hour.

AREA

Small and medium areas

square foot, square metre, square anything – As the names suggest, really. A square measuring a foot, or metre, or whatever else, on all four sides.

jo – The size of a Japanese tatami mat, six kanejaku (feet) by three.

tsubo – Two tatami mats, side by side, making up the six-foot square which is the basis of building and land sizing in Japan. Commonly used by Japanese farmers to describe the size of fields.

Bigger areas

are – A metric unit of a hundred square metres. Not much used nowadays.

acre – A unit for measuring fields in English-speaking countries. Like the French *journal*, it represents a day's work at the plough, and is a strip a furlong in length by a chain (see above) in width. Visualized as a squarer sort of shape, it's about half a soccer pitch.

hectare – A metric unit of area, but not an officially recognized SI one. It's a hundred ares, or one square hectometre. A hectometre is a hundred metres: this makes the hectare ten thousand square metres, or just under two and a half acres.

Acknowledgements

I would like to thank all those who have shared with me their personal tips and rules-of-thumb for how to measure things when you can't be bothered to lay your hands on the proper kit; and all those who helped with kind words, advice and encouragement during the writing of *About the Size of It* – in particular Alexander McCall Smith, Jilly Cooper, Conn Iggulden, Andrew Roberts, Sandy Gall, Edward Fox and Michael Trend.

I would also like to thank my good friend Vivian Linacre, whose book *The General Rule* offers a far more thorough and detailed account of customary measures than mine; and who is, besides, a great expert on megalithic measurements, which I have done less than justice to in this book.

Finally, I would like to express my admiration for the late Steven Thoburn, the Sunderland market trader who committed the 'crime' of selling a pound of bananas to an undercover trading-standards officer and who, with a handful of fellow small shopkeepers similarly charged, fought all the way to the High Court in London.

Photographic acknowledgements (by page)

Associated Press – *73*

Australian Environment Ministry (fossilized Ice Age footprint,
New South Wales) – *28*

The Bridgeman Art Library – *72, 109, 120*

Getty Images – *37, 124, 126*

Massachusetts Institute of Technology (MIT) – *59*

Shutterstock – *22, 23, 25, 31, 41, 48, 54*

www.panmacmillan.com

Printed in Great Britain
by Amazon

54546412R00097